Essential Biostatistics

A Nonmathematical Approach

HARVEY MOTULSKY
GraphPad Software, Inc.

New York Oxford
OXFORD UNIVERSITY PRESS

Oxford University Press is a department of the University of Oxford.
It furthers the University's objective of excellence in research,
scholarship, and education by publishing worldwide.

Oxford New York
Auckland Cape Town Dar es Salaam Hong Kong Karachi
Kuala Lumpur Madrid Melbourne Mexico City Nairobi
New Delhi Shanghai Taipei Toronto

With offices in
Argentina Austria Brazil Chile Czech Republic France Greece
Guatemala Hungary Italy Japan Poland Portugal Singapore
South Korea Switzerland Thailand Turkey Ukraine Vietnam

For titles covered by Section 112 of the US Higher Education
Opportunity Act, please visit www.oup.com/us/he for the
latest information about pricing and alternate formats.

Published by Oxford University Press
198 Madison Avenue, New York, New York 10016
http://www.oup.com

Library of Congress Cataloging-in-Publication Data
Motulsky, Harvey.
Essential biostatistics : a nonmathematical approach / Harvey Motulsky,
GraphPad Software, Inc.
pages cm
Includes bibliographical references and index.
ISBN 978-0-19-936506-7
1. Medicine--Research--Statistical methods. 2. Biometry. 3. Mathematical
statistics. I. Title.
R853.S7M682 2016
610.72--dc23

2015013963

BRIEF CONTENTS

CONTENTS

PREFACE

Essential Biostatistics: A Nonmathematical Approach is a concise and affordable nonmathematical guide to statistical thinking designed as a stand-alone short text, as a supplement to longer statistics textbooks that take a more mathematical approach, or as a review for scientists.

How is this Essentials edition different than the full *Intuitive Biostatistics*?
Intuitive Biostatistics was first published in 1995, and is now in its third edition. While many have praised the book, some teachers find it a bad fit for use in their course because it doesn't contain the mathematical content they need in their primary textbook, yet it is too long to serve as a supplemental text. To bridge this gap, I wrote *Essential Biostatistics: A Nonmathematical Approach*, which is about one-third the length of *Intuitive Biostatistics* (3rd edition).

Topics in *Intuitive Biostatistics* but missing from *Essential Biostatistics: A Nonmathematical Approach* include meta-analysis, survival curves, testing for equivalence or noninferiority, nonlinear and multiple regression, comparing fits of alternative models, and receiver operating characteristic (ROC) curves. Additionally, *Intuitive Biostatistics* (but not *Essential Biostatistics: A Nonmathematical Approach*) includes more than 100 pages explaining basic statistical tests and 40 pages of review problems with answers.

A unique approach
Some ways in which this book is unique:

- Chapter 1 is a fun chapter that explains how common sense can lead you astray and why we therefore need to understand statistical principles.
- Chapter 2 is a unique approach to appreciating the complexities of probability.
- I introduce statistical thinking with Chapter 4, which explains the confidence interval of a proportion. This lets me explain the logic of generalizing from sample to population using a confidence interval before having to deal with concepts about how to quantify the scatter. Don't make the

mistake of skipping Chapter 4 because you don't analyze data expressed as proportions, as this chapter explains some of the most important concepts in statistics.

- I explain comparing groups with confidence intervals (Chapter 12) before explaining P values (Chapter 13) and statistical significance (Chapters 14 and 15). This way I could delay as long as possible dealing with the confusing concept of a P value and the overused word "significant".
- Chapter 16 explains how common Type I errors are, and the difference between a significance level and the false discovery rate.
- Chapter 19 explains all common statistical tests as a series of tables.
- I include topics often omitted from introductory texts, but that I consider to be essential, including: multiple comparisons, the false discovery rate, p-hacking, lognormal distributions, geometric mean, normality tests, outliers and nonlinear regression.
- Nearly every chapter has a *Lingo* section explaining how statistical terminology can be misunderstood.
- To help you avoid making common misinterpretations, nearly every chapter includes a *Common Mistakes section*, and Chapter 25 explains more general mistakes to avoid.

Free GraphPad Web QuickCalcs and GraphPad InStat for iPad

QuickCalcs are free calculators I created (with help) on www.graphpad.com. Enter data and instantly get back results without saving or opening files. All of the calculators are either self-explanatory or link to longer explanations. If you need additional statistical calculators, check out the very long list of links maintained by John C. Pezzullo at www.statpages.org. As this book is going to press, GraphPad Software is creating a simple statistics program for the iPad. Search for GraphPad InStat on the App store.

Who helped?

A huge thanks to the many people listed below who reviewed draft chapters. Their contributions immensely improved this book:

Abdel-Salam G. Abdel-Salam, Virginia Polytechnic Institute and State University
B. Carol Adjemian, Pepperdine University
Raid Amin, University of West Florida
Michael Biro, Swarthmore College
Patrick Breheny, University of Kentucky
Michael F. Cassidy, Marymount University
Dean W. Coble, Stephen F. Austin State University
William M. Cook, Saint Cloud State University
Erica A. Corbett, Southeastern Oklahoma State University
Vincent A. DeBari, Seton Hall University
Bianca Frogner, George Washington University
Robert M. Hamer, University of North Carolina at Chapel Hill

Philip Hejduk, University of Texas at Arlington
Ravi P. Joshi, Old Dominion University
Stefan Judex, Stony Brook University
Chris Kerth, Texas A&M University
Joshua Lallaman, Saint Mary's University of Minnesota
Gary A. Lamberti, University of Notre Dame
Bruce D. Leopold, Mississippi State University
Susan P. McGorray, University of Florida
Matt McQueen, University of Colorado Boulder
Sumona Mondal, Clarkson University
Christopher J. Salice, Texas Tech University
David A. Sanchez, University of Texas at Arlington
Evelyn H. Schlenker, University of South Dakota
Andrew Jay Tierman, Saginaw Valley State University
Kathryn Trinkaus, Washington University
Derek Webb, Bemidji State University
Mary M. Whiteside, University of Texas at Arlington
John W. Wilson, University of Pittsburgh
Naji Younes, George Washington University

I also thank everyone at Oxford University Press who helped turn my manuscript into a polished book: Jason Noe, Senior Editor; Andrew Heaton, Assistant Editor; Patrick Lynch, Editorial Director; John Challice, Publisher and Vice President; Bill Marting, National Sales Manager; Frank Mortimer, Director of Marketing; David Jurman, Marketing Manager; Elizabeth Geist, Marketing Assistant; Lisa Grzan, Production Manager; Amy Whitmer, Production Team Leader; Christian Holdener, Senior Production Manager; Michele Laseau, Art Director; and Bonni Leon-Berman, Designer.

My relevant experience

For over a decade, I taught statistics to first-year medical students and/or biomedical sciences graduate students at the University of California, San Diego. The syllabus for those courses grew into the first edition of *Intuitive Biostatistics*, the original comprehensive text upon which this book is based. I no longer teach these courses, but as founder and CEO of GraphPad Software, I exchange emails with students and scientists almost daily, so am constantly reminded of the many ways that statistical concepts can be confusing or misunderstood.

Please email me with your comments, corrections, and suggestions for the next edition. I'll post errata at www.intuitivebiostatistics.com.

Harvey Motulsky
hmotulsky@graphpad.com

CHAPTER 1

Statistics and Probability Are Not Intuitive

Why do we need to use statistical methods? This chapter demonstrates how our instincts lead us astray when dealing with probabilities.

WE TEND TO JUMP TO CONCLUSIONS

A three-year-old girl told her male buddy, "You can't become a doctor; only girls can become doctors." To her, this statement made sense, because the three doctors she knew were all women.

When my oldest daughter was four, she "understood" that she was adopted from China, whereas her brother "came from Mommy's tummy." When we read her a book about a woman becoming pregnant and giving birth to a baby girl, her reaction was "That's silly. Girls don't come from Mommy's tummy. Girls come from China." With only one person in each group, she made a general conclusion. And when new data contradicted that conclusion, she—like many scientists—questioned the accuracy of the new data rather than the validity of her conclusion.

To avoid our natural inclination to make overly strong conclusions from limited data, scientists need to use statistics.

WE TEND TO BE OVERCONFIDENT

How good are you at knowing how sure you are? Find out using a test devised by Russo and Schoemaker (1989). Answer each of the questions below with a range that you think has a 90% chance of containing the correct answer. Don't use Google to find the answer. Don't give up and say you don't know. Of course you won't know the answers precisely! The goal is not to provide precise answers but to quantify your uncertainty and come up with ranges of values that you think are 90% likely to include the true answers. If you have no idea, answer with a really wide interval. For example, if you truly have no idea at all about the answer to the first question, answer with the range 0 to 120 years old, which you can be 100% sure includes the true answer. But try to narrow your responses to a range that you are 90% sure contains the right answer:

- Age of Martin Luther King, Jr., at his death
- Length of the Nile River, in miles or kilometers
- Number of countries in OPEC
- Number of books in the Old Testament
- Diameter of the moon, in miles or kilometers
- Weight of an empty Boeing 747, in pounds or kilograms
- Year when Mozart was born
- Gestation period of an Asian elephant, in days
- Distance from London to Tokyo, in miles or kilometers
- Deepest known point in the ocean, in miles or kilometers

Compare your answers with the correct answers listed on page 5. If you were 90% sure of each answer, you would expect about nine intervals to include the correct answer and one interval to exclude it.

Russo and Schoemaker (1989) tested more than 1,000 people and reported that 99% of them were overconfident. The goal was to create ranges that included the correct answer 90% of the time, but most people created narrow ranges that included the correct answers only 30% to 60% of the time. Studies of experts estimating facts in their areas of expertise led to similar results.

Since we tend to be too sure of ourselves, scientists must use statistical methods to quantify confidence properly.

WE SEE PATTERNS IN RANDOM DATA

Table 1.1 presents simulated data from 10 basketball players (1 per row) shooting 32 baskets each. An "X" represents a successful shot, and a "–" represents a miss. Is this pattern random? Or does it show nonrandom streaks of successful shots?

Most people see patterns, yet Table 1.1 was generated randomly. Each spot had a 50% chance of being an "X" (a successful shot) and a 50% chance of being a "–" (an unsuccessful shot), without taking into consideration previous shots. So why do we see clusters? Perhaps because our brains have evolved to find patterns and do so very well. This ability may have served our ancestors to avoid predators and poisonous plants, but it is important that we recognize this built-in mental bias. Statistical rigor is essential to avoid being fooled by seeing apparent patterns among random data.

WE DON'T EXPECT VARIABILITY TO DEPEND ON SAMPLE SIZE

Gelman (1998) looked at the relationship between the populations of counties and the age-adjusted, per-capita incidence of kidney cancer (a fairly rare cancer, with an incidence of about 15 cases per 100,000 adults in the United States). First, he focused on counties with the lowest per-capita incidence of kidney cancer. Most

–	–	X	–	X	–	–	X	X	–	–	–	–	X	X	X	–	X	X	–	X	X	–	–	–	X	X	–	–	–	X	X
X	–	–	X	–	X	X	–	–	X	X	–	–	X	–	X	–	X	–	–	–	X	X	X	X	–	–	X	X	–	–	–
X	X	X	X	–	X	X	–	X	–	X	–	X	X	X	–	–	–	–	–	X	–	X	–	X	X	X	–	–	–	–	X
–	X	–	X	–	–	X	X	–	X	X	–	X	X	–	–	X	X	X	X	–	–	–	–	X	X	–	X	–	X	–	–
–	X	–	X	–	X	X	–	–	–	X	X	–	–	–	–	–	–	X	–	X	–	X	–	–	X	–	–	X	–	X	X
–	–	X	X	X	–	X	–	X	–	–	–	X	X	X	X	–	X	X	X	X	–	–	–	–	–	X	X	–	X	X	X
X	–	–	X	X	–	–	X	X	X	X	–	X	X	X	–	–	X	–	–	X	X	X	X	X	–	X	X	X	–	–	–
X	–	X	–	–	–	X	X	X	X	X	–	–	X	X	–	X	X	–	X	X	X	–	X	X	–	X	–	–	X	–	X
X	X	X	–	–	X	X	X	X	X	–	X	–	X	–	X	X	–	X	–	X	X	X	X	–	X	X	–	X	X	X	X
–	–	–	X	X	X	–	–	X	X	X	–	X	X	X	–	–	X	–	–	X	–	X	X	X	X	X	–	–	–	X	–

Table 1.1. Random patterns don't seem random.

This table represents 10 basketball players (1 per row) shooting 32 baskets each. An "X" represents a successful shot, and a "–" represents a miss. Is this pattern random? Or does it show signs of nonrandom streaks? Most people tend to see patterns, but in fact, the arrangement is entirely random. Each spot in the table had a 50% chance of having an "X."

of these counties had small populations. Why? One might imagine that something about the environment in these rural counties leads to lower rates of kidney cancer. Then, he focused on counties with the highest incidence of kidney cancer. These also tended to be the smallest counties. Why? One might imagine that lack of medical care in these tiny counties leads to higher rates of kidney cancer. But it seems pretty strange that both the highest and lowest incidences of kidney cancer be in counties with small populations?

The reason is simple, once you think about it. In large counties, there is little variation around the average rate. Among small counties, however, there is much more variability. Consider an extreme example of a tiny county with only 1,000 residents. If no one in that county had kidney cancer, that county would be among those with the lowest (zero) incidence of kidney cancer. But if only one of those people had kidney cancer, that county would then be among those with the highest rate of kidney cancer. In a really tiny county, it only takes one case of kidney cancer to flip from having one of the lowest rates to having one of the highest rates. In general, just by chance, the incidence rates will vary much more among counties with tiny populations than among counties with large populations. Therefore, counties with both the highest and the lowest incidences of kidney cancer tend to have smaller populations than counties with average incidences of kidney cancer.

There is more random variation within small groups than within large groups. This simple principle is logical, yet is not intuitive to many people.

WE ARE FOOLED BY MULTIPLE COMPARISONS

Austin and colleagues (2006) sifted through a database of health statistics for 10 million residents of Ontario, Canada. They examined 223 different reasons for hospital admission and recorded the astrological sign of each patient (computed from his or her birth date). They then asked if people with certain astrological signs are more likely to be admitted to the hospital for certain conditions.

The results seem impressive. The incidence of seventy-two conditions was statistically significantly higher in people with one particular astrological sign than in people with all the other astrological signs put together. Essentially, a result is "statistically significant" when it would occur by chance less than 5% of the time (you'll learn more about what "statistically significant" means in Chapter 14).

Sounds impressive, doesn't it? Indeed, the data make you think there is a convincing relationship between astrology and health. But there is a problem. It is misleading to focus on the strong associations between one disease and one astrological sign without considering all the other combinations. By testing the association of 223 different medical conditions with 12 astrological signs, Austin and colleagues (2006) actually tested 2,676 distinct hypotheses ($223 \times 12 = 2{,}676$). Therefore, they would expect to find 134 statistically significant associations just by random chance (5% of 2,676 = 134) but in fact only found 72.

Note that this study wasn't really done to ask about the association between astrological sign and disease. It was done to demonstrate the difficulty of interpreting statistical results when scientists make many comparisons. (Chapter 17 explores multiple comparisons in more depth.)

WE TEND TO IGNORE ALTERNATIVE EXPLANATIONS

Imagine you are studying the use of acupuncture in treating osteoarthritis. Patients who come in with severe arthritis pain are treated with acupuncture and asked to rate their arthritis pain before and after the treatment. The pain decreases in most patients after treatment, and statistical calculations show that such consistent findings are exceedingly unlikely to happen by chance. Therefore, the acupuncture must have worked. Right?

Not necessarily. The decrease in recorded pain may not be caused by the acupuncture. Here are five alternative explanations (adapted from Bausell, 2007):

- If the patients believe in the therapist and treatment, that belief may reduce the pain considerably. The pain relief may be a placebo effect and have nothing to do with the acupuncture itself.
- The patients want to be polite and may tell the experimenter what he or she wants to hear (that the pain decreased). Thus, the decrease in reported pain may be because the patients are not accurately reporting pain after therapy.
- Before, during, and after the acupuncture treatment, the therapist talks with the patients. Perhaps he or she recommends a change in aspirin dose, a change in exercise, or the use of nutritional supplements. The decrease in reported pain might be due to these aspects of the treatment rather than the acupuncture.
- The experimenter may have altered the data. For instance, what if three patients experience worse pain with acupuncture, whereas the others get better? The experimenter carefully reviews the records of those three

patients and decides to remove them from the study. One of those people actually has a different kind of arthritis than the others, and two had to climb stairs to get to the appointment because the elevator didn't work that day. The data, then, are biased or skewed because of the omission of these three participants.
- The pain from osteoarthritis varies significantly from day to day. People tend to seek therapy when pain is at its worst. If you start keeping track of pain on the day when it is the worst, it is quite likely to get better even without treatment.

WE CRAVE CRISP CONCLUSIONS, BUT STATISTICS OFFERS PROBABILITIES

Many people expect statistical calculations to yield definitive conclusions. But in fact, every statistical conclusion is stated in terms of probability. Statistics can be very difficult to learn if you keep looking for definitive conclusions. As statistician Myles Hollander reportedly said, "Statistics means never having to say you're certain!"

CHAPTER SUMMARY

- Our brains do a really bad job of interpreting data. We see patterns in random data, tend to be overconfident in our conclusions, and mangle interpretations that involve combining probabilities.
- Our intuitions tend to lead us astray when interpreting probabilities or multiple comparisons.
- We need statistical (and scientific) rigor to avoid reaching invalid conclusions.

Answers to the 10 questions in the "We Tend to be Overconfident" section.

Age of Martin Luther King, Jr., at his death: 39
Length of the Nile River: 4,187 miles or 6,738 kilometers
Number of countries in OPEC: 13
Number of books in the Old Testament: 39
Diameter of the moon: 2,160 miles or 3,476 kilometers
Weight of an empty Boeing 747: 390,000 pounds or 176,901 kilograms
Year when Mozart was born: 1756
Gestation period of an Asian elephant: 645 days
Distance from London to Tokyo: 5,989 miles or 9,638 kilometers
Deepest known point in the ocean: 6.9 miles or 11.0 kilometers

CHAPTER 2

The Complexities of Probability

BASICS OF PROBABILITY

Probabilities range from 0.0 to 1.0 and are used to quantify a prediction about future events or the certainty of a belief:

- A probability of 0.0 (or 0%) means that an event can't happen, or that someone is absolutely sure a statement is wrong.
- A probability of 1.0 (or 100%) means that an event is certain to happen, or that someone is absolutely certain a statement is correct.
- A probability of 0.50 (or 50%) means that an event is equally likely to happen or not happen, or that someone believes a statement is equally likely to be true or false.

PROBABILITY AS PREDICTION OF LONG-TERM FREQUENCY

Probabilities as predictions from a model

This chapter focuses on one simple example: a woman plans to get pregnant and wants to know the chance that her baby will be a boy. One way to think about probability is as the prediction of a future event that we derive by using a model. A *model* is a simplified description of a mechanism. For this example, here is a simple model:

- Each ovum has an X chromosome, and none has a Y chromosome.
- Half the sperm have an X chromosome (but no Y), and half have a Y chromosome (but no X).
- Only one sperm will fertilize the ovum.
- Each sperm has an equal chance of fertilizing the ovum.
- If the winning sperm has a Y chromosome, the fetus will have both an X and a Y chromosome and so will be male. If the winning sperm has an X chromosome, the fetus will have two X chromosomes and so will be female.
- Any miscarriage or abortion is equally likely to happen to a male or a female fetus.

If you assume that this model is true, then the predictions are easy to figure out. Since all sperm have the same chance to fertilize the ovum, and since the sperm are equally divided between those with an X chromosome and those with a Y chromosome, the chance that the fetus will have a Y chromosome is 50%. Thus, our model predicts that the chance the fetus will be male is 0.50, or 50%. (In any particular group of babies, the fraction of boys might be more or less than 50%. But in the long run, you'll expect 50% of the babies to be boys.)

Probabilities based on data

Of all babies born in the world in 2011, 51.7% were boys (Central Intelligence Agency, 2012). We don't need to know *why* more boys than girls were born. It is simply a fact that this sex ratio is imbalanced in many countries. Based on these data, we can say that the chance a baby will be a boy is 51.7%. If you had a group of 1,000 women giving birth, you'd expect about 517 to have male babies and 483 to have female babies. (In any particular set of 1,000 women giving birth, the number of males might be higher or lower than 517, but again, this is what you'd expect in the long run.)

PROBABILITY AS STRENGTH OF BELIEF (BAYES)

Subjective probabilities

You badly want a boy. You search the Internet and read about a book that explains a method said to give couples a 75% or better chance of having a boy. The reviews of this method are glowing and convince you that its premise is correct. Therefore, you plan to follow the book's recommendations to increase your chance of having a boy.

What is the chance that you'll have a boy? If you have complete faith that the method is correct, then you believe that the probability, as stated on the book jacket, is 75%. You don't have any data supporting this number, but you believe it strongly. It is your firmly held *subjective probability*.

Of course, different people have different beliefs about the efficacy of the method. There are no studies actually proving that the method works, and one (in a prestigious journal) says that it doesn't. So I think your chance of having a boy is 51.7% (the same as if you hadn't used that method).

Since you and I have different assessments regarding the probability of whether the method works, my answer for the chance of you having a boy doesn't match your answer. A large field of statistics, called *Bayesian statistics*, uses this definition of probability.

Probability as quantification of ignorance

Assume that a woman is pregnant but hasn't yet done any investigations to determine whether the fetus is male or female. Nor has she undergone any interventions that purport to change the chances of having a boy or a girl. What is the probability that the fetus is male?

In one sense, the concept of probability is simply not involved. The random event was the race of many sperm to fertilize the egg. One sperm won, and the fetus either is male or is female. Since that random event has already happened and the sex of the fetus is now a fact, you could argue that it makes no sense to ask about probability or chance or odds. The issue is ignorance, not randomness.

Another perspective: Before she became pregnant, there was a 51.7% chance that her baby would be male. If many thousands of women were pregnant without knowing the sex of their fetus, you'd expect about 51.7% to have male babies. Therefore, it seems sensible to say that there is about a 51.7% chance that the baby will be male and a 48.3% chance that it will be female. But notice the change in perspective. Now, you are no longer talking about predicting the outcome of a random future event. Instead, you are quantifying your ignorance of an event that has already happened.

THE DISTINCTION BETWEEN PROBABILITY AND STATISTICS

The words *probability* and *statistics* are often linked together in course and book titles, but they are distinct and work in opposite directions (Table 2.1).

Probability calculations start with the general case, called the *population* or *model*, and predict what will happen in many samples of data. Probability calculations go from general to specific, from population to sample, and from model to data.

Statistical calculations start with one set of data (the *sample*) and make inferences about the overall population or model. The logic goes from specific to general, from sample to population, and from data to model.

LINGO

If you search for demographic information on the fraction of babies who are boys, most likely you'll find the *sex ratio*. This term, as used by demographers, is the ratio of males to females born. Worldwide, the sex ratio at birth in many

PROBABILITY		
General	→	Specific
Population	→	Sample
Model	→	Data
STATISTICS		
General	←	Specific
Population	←	Sample
Model	←	Data

Table 2.1. The distinction between probability and statistics.

countries is about 1.07. Equivalently, the odds of having a boy versus a girl are 1.07 to 1.00, or 107 to 100.

Odds and *probability* are two alternative ways to express precisely the same concept. Every probability can be expressed as odds. Any chance expressed as odds can also be expressed as a probability. Some scientific fields tend to prefer using probabilities; others tend to favor odds. There is no consistent advantage to using one or the other.

It is easy to convert the odds to a probability. If there are 107 boys born for every 100 girls born, the chance that any particular baby will be male is 107/(107 + 100) = 0.517, or 51.7%.

It is also easy to convert from a probability to odds. If the probability of having a boy is 51.7%, then for every 1,000 births you'd expect 517 boys and 483 girls (1,000 − 517 = 483) to be born. So the odds of having a boy versus a girl are 517/483 = 1.07 to 1.00. The odds are defined as the probability that the event will occur divided by the probability that the event will not occur.

Odds can be any positive value or zero, but they cannot be negative. A probability must range from 0.0 to 1.0 if expressed as a fraction or from 0 to 100 if expressed as a percentage.

The probability of flipping a coin to heads is 50%. The odds are 50:50, which equals 1.0. So a probability of 0.5 is the same as odds of 1.0, or 1 to 1. As the probability goes from 0.5 to 1.0, the odds increase from 1.0 with no limit. For example, if the probability is 0.75, then the odds are 75:25, or 3 to 1, or 3.0. If the probability is 0.99, then the odds are 99:1, 99 to 1, or 99.0.

COMMON MISTAKES

Mistake: Ignoring the assumptions

The examples above asked, "What is the chance that a baby will be male?" However, that question is meaningless without context. For it to be meaningful, you must accept a list of assumptions, including:

- We are asking about human babies. Sex ratios may be different in other species.
- There is only one baby. The question needs elaboration if you allow the possibility of twins or triplets.
- There is only a tiny probability (ignored here) that the baby is neither completely female nor completely male.
- The sex ratio is the same for all countries and all ethnic groups.
- The sex ratio does not change from year to year, or between seasons.
- There will be no sex-selective abortions or miscarriages, so the sex ratio at conception is the same as the sex ratio at birth.

Whenever you hear questions or statements about probability, remember that probabilities are *always* contingent on a set of assumptions. So to think clearly about probability in any situation, you must know what those assumptions are.

Mistake: Reversing probability statements

With the examples we've used so far, there is no danger of accidentally reversing probability statements by mistake. The probability that a baby is a boy is obviously very different than the probability that a boy is a baby. But in many situations, it is easy to get things backward. For example:

- The probability that a heroin addict first used marijuana is not the same as the probability that a marijuana user will later become addicted to heroin.
- The probability that someone with abdominal pain has appendicitis is not the same as the probability that someone diagnosed with appendicitis will have abdominal pain.
- The probability that a statistics books will be boring is not the same as the probability that a boring book is about statistics.

Mistake: Believing that probability has a memory

If a couple has four children, all boys, what is the chance that their next child will be a boy? A common mistake is to think that since this couple already has four boys, they are somehow due to have a girl, so the chance of having a girl is elevated. This is simply wrong.

This mistake is often made in gambling, especially when playing roulette or a lottery. Some people bet on a number that has not come up for a long while, believing that there is an elevated chance of that number coming up on the next spin. This is called the *gambler's fallacy*. Probability does not have a memory.

But if probability does not have a memory, how do the probabilities become more stable with large sample sizes? For example, if you flip lots of coins, early on there may be more heads than tails. Over time, however, the number of heads and the number of tails will eventually become very similar. This is because the imbalance becomes diluted as the number of trials goes up, not because the imbalance is corrected due to an increase in the chance of the coin ending up tails.

CHAPTER SUMMARY

- Probability can be a confusing topic.
- The term 'probability' has two meanings. One is the long-term chance that an event will happen. The other is quantifying how sure one is about the truth of a proposition.
- All probability statements are based on a set of assumptions.
- Since a probability is essentially a fraction, it is always essential to clearly define both its numerator and its denominator.
- Probability calculations go from general to specific, from population to sample, and from model to data. Statistical calculations work in the opposite direction: from specific to general, from sample to population, and from data to model.

CHAPTER 3

From Sample to Population

SAMPLING FROM A POPULATION

The distinction between a sample and a population is key to understanding much of statistics. Statisticians say that you analyze data collected from a sample to make conclusions about the population from which the data were sampled. Note that the terms *sample* and *population* have specialized meanings in statistics that differ from the ordinary uses of those words. As you learn statistics, distinguish the specialized definitions of statistical terms from their ordinary meanings.

Here are four of the different contexts in which these terms are used:

- Quality control. A factory makes lots of items (the population) but randomly selects a few of those items to test (the sample). The results obtained from the sample are used to make inferences about the entire population.
- Political polls. A random sample of voters is polled, and the results are used to make conclusions about the entire population of voters.
- Clinical studies. Data from the sample of patients included in a study are used to make conclusions useful to a larger population of future patients.
- Laboratory experiments. The data you collect are the sample. From the sample data, you want to make reliable inferences about the ideal, true underlying situation (the population).

In quality control as well as in political (and marketing) polls, the population is usually much larger than the sample, but it is finite and known (at least approximately). In biomedical research, we usually *assume* that the population is infinite, or at least very large compared with our sample. Standard statistical methods assume that the population is much larger than the sample. If the population has a defined size and you have sampled a substantial fraction of it (>10% or so), then you must use special statistical methods designed for when a large fraction of the population is sampled. These are beyond the scope of this book.

HOW FAR TO GENERALIZE?

Roberts (2004) published 10 different self-experiments. For example, Roberts found that when he drank unflavored fructose water between meals, he not only lost weight but also maintained the lower weight for more than a year. Statistical analysis of the sample of data he collected enabled him to make conclusions that he found helpful for managing his life. However, the statistical methods only let him generalize to what would have happened had he collected many more measurements in his experiment. Statistical methods cannot be used to make inferences about what would happen if others drank unflavored fructose water between meals. His data present an intriguing idea for a broader experiment, but you can't use data from one person to make statistical inferences about others.

Another example: Imagine you are working on the physiology of heart failure. You induce heart failure in some animals and then compare the heart tissue from animals with and without heart failure. If the animals are small (e.g., mice), you might collect tissue from several hearts in each group, then compare the two pooled preparations with biochemical assays repeated a few times. Statistical methods let you make inferences about what might have happened had you repeated the measurements from each tissue preparation many times, which in turn lets you make conclusions about those particular animals. However, statistical calculations using these data do not let you make broader conclusions about other animals or other species (e.g., human). Broader conclusions require common sense and scientific judgment, and statistical principles don't help.

LINGO

Sampling error and bias

Much of statistics is based on the assumption that the data you are analyzing are randomly sampled from a larger population. Therefore, the values you compute from the sample (e.g., the mean, the proportion of cases that have a particular attribute, or the best-fit slope of a linear regression line) are considered to be estimates of the true population values. Again, notice that a common word ("estimate") is given a special meaning in statistics.

There are several reasons why a value computed from a sample differs from the true population value:

- *Sampling error.* Just by chance, the sample you collected might have a higher (or lower) mean than that of the population. Just by chance, the proportion computed from the sample might be higher (or lower) than the true population value. Just by chance, the best-fit slope of a regression line (which you will learn about in Chapter 23) might be steeper (or shallower) than the true line that defines the population.
- *Selection bias.* The difference between the value computed from your sample and the true population value might be due to more than random

sampling. It is possible that the way you collected the sample is not random but instead preferentially samples subjects with particular values. For example, political polls are usually done by telephone and thus select for people who own a phone (and especially those with several phone lines) and for people who answer the phone and don't hang up on pollsters. These people may have opinions very different than those in a true random sample of voters.

- Other forms of *bias*. The experimental methodology may be imperfect and give results that are systematically either too high or too low. Bias is not random but rather produces errors in a consistent direction.

Models and parameters

The concept of sampling from a population is an easy way to understand the idea of making generalizations from limited data, but it doesn't always perfectly describe what statistics and science are all about. Another way to think about statistics is in terms of models and parameters.

A *model* is a mathematical description of a simplified view of the world. A model consists of both a general description and *parameters* that have particular values. One goal of statistics is to analyze your data to make inferences about the values of the parameters that define the model. All statistical calculations are based on assumptions, and a model clarifies what these are.

For example, the model might describe temperature values that follow a Gaussian bell-shaped distribution (see Chapter 8). The parameters of that distribution are its mean (average) and standard deviation (a measure of variation; see Chapter 7).

Another example: After a person ingests a drug, its concentration in the blood may decrease according to an exponential decay with a consistent half-life. The parameter is the half-life. In one half-life, the drug concentration goes down to 50% of the starting value; in two half-lives, it goes down to 25%; in three half-lives, it goes down to 12.5%; and so on. A statistical analysis, nonlinear regression (see Chapter 24), can fit the model to the concentration data to determine the value of the half-life.

Random samples versus convenience samples

In clinical studies, it is not feasible to randomly select patients from the entire population of similar patients. Instead, patients are selected for the study because they happen to be at the right clinic at the right time. This is called a *convenience sample* rather than a *random sample*. For statistical calculations to be meaningful, we must assume that the convenience sample adequately represents the population and that the results are similar to what would have been observed had we used a true random sample.

Some kinds of experiments use *systematic sampling* as a reasonable alternative to random sampling. For example, study every fourth patient who comes to a clinic, or dig up a soil sample every 100 feet along a line.

COMMON MISTAKES

Mistake: Using statistical methods when the data are complete so there is no extrapolation

Statistical methods let you make general inferences about a population after analyzing a sample of data. It is a mistake to use statistical methods when there is no need for statistical inference.

Say you are a professor and want to compare the scores on a final exam this year with the scores on the same exam last year. You have all the scores, and you can ask which has the highest average and how much the two distributions overlap. There is no need to do any statistical calculations (beyond descriptive statistics), because you are not making an inference about larger populations. You have data for every student in both years, and you simply want to compare those two data sets. Since you are not trying to make conclusions about a broader population, there is no need for statistical inference.

Mistake: Thinking that if your sample size is large, bias doesn't matter

Having a large sample size can help if you have an unbiased sampling method. But if your sampling method is biased, collecting a larger sampling method won't get you more accurate results.

CHAPTER SUMMARY

- The goal of data analysis is simple: to make the strongest possible conclusions from limited amounts of data.
- Statistics help you generalize from a particular set of data (your sample) to make a more general conclusion (about the population). Statistical calculations go from specific to general, from sample to population, from data to model.
- Bias occurs when the experimental design is faulty so that on average the result is too high or too low. Results are biased when the result computed from the sample is not in fact the best possible estimate of the value in the population.
- The results of statistical calculations are always expressed in terms of probability.

CHAPTER 4

Confidence Intervals

Confidence intervals express precision or margin of error and so let you make a general conclusion from limited data. This chapter focuses on confidence intervals of proportions. But even if you only work with continuous data rather than proportions, this is an essential chapter because it explains key concepts in statistics.

EXAMPLE: SURVIVAL OF PREMATURE INFANTS

To better counsel the parents of premature babies, Allen, Donohue, and Dusman (1993) investigated the survival of premature infants. They retrospectively studied all premature babies born at 22 to 25 weeks of gestation at their hospital during a three-year period. Of 29 infants born at 22 weeks of gestation, none survived past six months. Of 39 infants born at 25 weeks of gestation, 31 (79.5%) survived for at least six months.

Is it worth doing any calculations to make more general conclusions? Only if you think it is reasonable to assume that these infants are somewhat representative of a larger population of premature babies born at that time in similar hospitals and you wish to make inferences about that larger group.

If you collected data on a much larger population of babies, the percentage of survival in the two groups would probably not be exactly 0% and 79.5%. But how far away from those numbers are the percentages likely to be? If you want to create a range of values that you are absolutely certain will contain the overall percentage from the much larger hypothetical population, it would have to run all the way from 0% to 100%. That wouldn't be useful. So let's create ranges of percentages that we are 95% sure will contain the true population values for the two groups of babies. Such a range is called a *confidence interval* (abbreviated CI). Confidence intervals will be more rigorously defined later in the chapter.

Before reading on, write down what you think the confidence intervals are for the two examples (0/29 and 31/39). Now, calculate the answer using the web calculator at www.graphpad.com/quickcalcs/confInterval1. This calculator computes the confidence intervals in two ways (with similar results). Using the modified Wald method, the 95% CI of 31/39 ranges from 64.2% to 89.5%, and the 95% CI of 0/29 extends from 0% to 13.9%. We can be 95% sure that these ranges include the overall proportion of surviving infants in the two populations.

These confidence intervals only account for sampling variability, also called *sampling error*. If you wanted to apply these results to another hospital, you also must account for the different populations served by different hospitals and the different methods used to care for premature infants. In addition, note that these data are two decades old, and the survival of premature infants has improved since then.

EXAMPLE: POLLING VOTERS

Say you polled 100 randomly selected voters just before an election, and 33 said they would vote for your candidate. What can you say about the proportion of all voters in the population who will vote for your candidate?

We'll assume that the sample is perfectly representative of the population of voters and that all people will vote as they said they would when polled. Even so, just by chance, your 100-voter sample will almost certainly contain either a smaller or a larger fraction of people voting for your candidate than the fraction in the overall population. In other words, we need to think about *sampling error*.

Because we only know the proportion of favorably inclined voters in one sample, there is no way to be sure about the proportion in the overall population. The best we can do is to calculate a range of values that we expect will span the true population proportion. How wide does this range of values have to be? Again, if we wanted to be 100% sure the range spans the true value, the range would have to run from 0% to 100%. And again, that isn't helpful. Instead, we create a 95% CI and thus accept a 5% chance the range will not include the true population value.

Before reading on, make your best guess for the 95% CI when your sample has 100 people and 33 of those have said they would vote for your candidate (so the proportion in the sample is 0.33, or 33%). Note that there is no uncertainty about what we observed in the voter sample. We are absolutely sure that 33.0% of the 100 people polled said they would vote for our candidate. If we weren't sure of that count, calculation of a confidence interval could not overcome any mistakes that were made in tabulating those numbers or any ambiguity in defining "vote." What we don't know is the proportion of favorable voters in the entire population. Write down your guess before continuing.

Lots of programs and web calculators can do the math for you. Here, the 95% CI extends from 25% to 43% (you might get slightly different results, depending on which program you use). So we can be 95% confident that somewhere between 25% and 43% of the population you polled preferred your candidate on the day the poll was conducted. The phrase "95% confident" is explained in more detail below.

Don't get confused by the two different uses of percentages. The confidence interval is for 95% confidence. That quantifies how sure you want to be. You could ask for a 99% CI if you wanted your range to have a higher chance of including the true value. Each confidence limit here (25% and 43%) represents the percentage of voters who will vote for your candidate.

ASSUMPTIONS: CONFIDENCE INTERVAL OF A PROPORTION

The whole idea of a confidence interval is to make a general conclusion from some specific data. This generalization is only useful if certain assumptions are true. A huge part of mastering statistics is learning the assumptions upon which statistical inferences are based.

Assumption: Random (or representative) sample

The 95% CI is only strictly valid when the sample of data you are analyzing was randomly drawn from a larger population about which you want to generalize. In other words, you have collected a *random sample.*

This assumption would be violated in the premature baby example if the infants in the sample were somehow selected to be more (or less) sick than other premature infants of similar age. This assumption would be violated in the election example if the sample was not randomly selected from the population of voters. In fact, this very mistake was made in the 1936 Roosevelt–Landon U.S. presidential election. To find potential voters, the pollsters relied heavily on phone books and automobile registration lists. But in 1936, Republicans were far more likely than Democrats to own a phone or a car, so the poll selected too many Republicans. There also was a second problem. For some reason, supporters of Landon were much more likely to return replies to the poll than were supporters of Roosevelt (Squire, 1988). The poll predicted that Landon would win by a large margin, but in fact, Roosevelt won.

Assumption: Independent observations

The 95% CI is only valid when all subjects are sampled from the same population and each has been selected independently of the others.

In the premature infant example, the data would not be independent if some of the infants in the sample were twins, because they would share genetic or environmental factors that might influence survival, or if some of the deaths were caused by an infection or malfunctioning equipment that affected several infants in the study. The assumption would be violated in the election example if the pollsters questioned more than one person in some families (because they would tend to vote the same way) or if some voters were polled more than once.

Assumption: Data were tabulated correctly

The 95% CI is only valid when the number of subjects in each category is tabulated correctly.

In the premature infant example, the outcome is death, which is pretty unambiguous. The data would not be accurate in the election example if the pollster deliberately recorded some of the opinions incorrectly or somehow coerced or intimidated the respondents to answer a certain way.

Results are said to be *biased* if the experimental methods or analysis procedures ensure that the calculated results, on average, deviate from the true value.

WHAT DOES 95% CONFIDENCE REALLY MEAN?

The true population value either lies within the 95% CI you calculated or it doesn't. There is no way for you to know. If you were to calculate a 95% CI from many samples, you would expect it to include the population proportion in about 95% of the samples and not to include the population proportion in about 5% of the samples.

A simulation

To demonstrate this, let's switch to a *simulation* in which you know *exactly* the population from which the data were selected. Assume that you have a bowl of 100 balls, 25 of which are red and 75 of which are black. Now pretend you are an experimenter who doesn't know what fraction of the balls are red. Mix the balls well, and choose one randomly. Put it back in, mix again, and choose another one. Repeat this process until you have chosen 15 times, record the fraction of the balls you drew that were red, and compute the 95% CI for that proportion.

Figure 4.1 shows the simulation of 20 such 15-ball samples. Each 95% CI is shown as a bar extending from the lower confidence limit to the upper confidence limit. The value of the observed proportion in each sample is shown as a line in the middle of each CI. The horizontal line shows the true population value (25% of the balls are red). In about half of the samples, the sample proportion is less than 25%, and in the other half, the sample proportion is higher than 25%. In the ninth sample, the population value lies outside the 95% CI. In the long run, given the assumptions listed earlier, 1 of 20 (5%) of the 95% CIs will not include the population value, and that is exactly what we see in this set of 20 simulated experiments.

Figure 4.1 helps explain confidence intervals, but you cannot create such a figure when you analyze data. This is because when you analyze data, you don't know the actual population value. You only have results from a single experiment. In the long run, 95% of such intervals will contain the population value, and 5% will not. But there is no way you can know whether a particular 95% CI includes the population value or not, because you don't know that population value (except when running simulations).

95% chance of what?

Assume that you repeat an experiment many times, that you haven't violated any of the assumptions, and that you compute a 95% CI for each experiment. In the long run, you expect 95% of these intervals to include the true population value, and the other 5% to not include that value. That is the definition of a 95% CI.

But when analyzing data, you have just one confidence interval. How do you interpret it? Many people (including me) say that there is a 95% chance that the

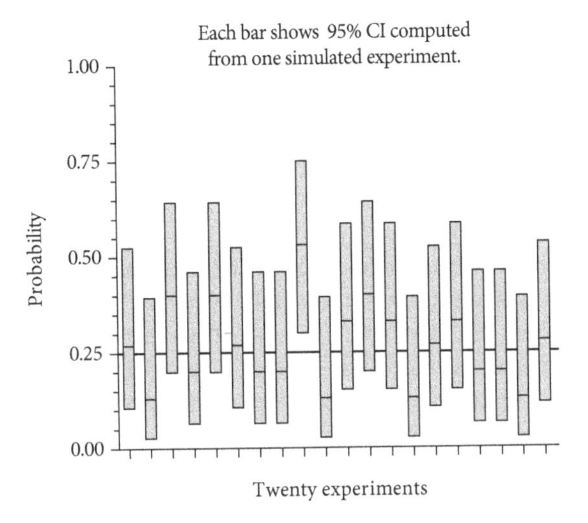

Figure 4.1. What would happen if you collected many samples and computed a 95% CI for each?

In all but one of the simulated samples, the confidence interval includes the true population proportion of red balls (shown as a horizontal line). The 95% CI of sample 9, however, does not include the true population value. You expect this to happen in 5% of samples. Because this figure shows the results of simulations, we know when the confidence interval doesn't include the true population value. When analyzing data, however, the population value is unknown, so you have no way of knowing whether or not the confidence interval includes the true population value.

95% CI you calculated contains the true population value. Others say that the interval either contains the true value or it doesn't. Once you have collected the data, chance is no longer involved, so these people object to the interpretation that a particular interval has a 95% chance of including the population value.

This is similar to the baby gender example in Chapter 2. If a woman is pregnant, but has done nothing to learn the sex of her fetus, what are the chances? I'd say there is a bit more than a 50% chance of it being a boy (see Chapter 2). Others would say that the fetus either is a boy or is a girl with no chance involved, so it makes no sense to talk about probabilities.

Back to confidence intervals. It is definitely a mistake to flip things around and say that there is a 95% chance that the population value lies within the calculated confidence interval. Random chance affects which data you collect, which in turn affects the range of the confidence interval you calculate. If you repeat the experiment, you'll almost certainly get a different interval. But random chance does not affect the true population value, which is fixed (but unknowable). Therefore, it is considered incorrect to say "There is a 95% chance that the population value is within this 95% CI." That statement implies that the population value is subject to random variation. Instead, most people consider it proper to say "There is a 95% chance that this 95% CI contains the population value." That statement correctly implies that the confidence interval is subject to random variation (if you were to repeat the experiment). Yes, this is a subtle distinction!

It would also be a mistake to say that if we were to repeat the experiment over and over, then 95% of the experiments would lead to parameters that lie within the confidence interval computed from the first experiment.

What is special about 95%?

Confidence intervals can be computed for any degree of confidence. By convention, 95% CIs are presented most commonly, although 90% and 99% CIs are sometimes published. With a 90% CI, there is a 10% chance that the interval does not include the population value. With a 99% CI, that chance is only 1%.

If you are willing to be less confident, then the interval will be *shorter*, so 90% CIs are narrower than 95% CIs. If you want to be more confident that your interval contains the true population value, the interval must be *longer*. Thus, 99% CIs are wider than 95% CIs. If you want to be 100% confident that the interval contains the population value, the interval needs to be wide enough to include every possible value, so this is not done.

ARE YOU QUANTIFYING THE EVENT YOU REALLY CARE ABOUT?

The 95% CI allows you to generalize from the sample to the population for the event that you tabulated. For the first example (premature babies), the tabulated event (death in the first six months of life) is exactly what you care about. But sometimes the event you really care about is different from the one you actually tabulated.

In the voting example, we assessed our sample's responses to a poll conducted on a particular date, so the 95% CI lets us generalize those results to predict how the population would respond on that date. We wish to extrapolate to election results in the future but can do so only by making an additional assumption: that people will vote as they said they would. Before the 1948 Dewey–Truman U.S. presidential election, polls of many thousands of voters indicated that Dewey would win by a large margin. Because the confidence interval was so narrow, the pollsters were very confident. Newspapers were so sure of the results that they even prematurely printed the headline "Dewey Beats Truman." In fact, Truman won. Why were the polls inaccurate? They were performed in September and early October, and the election was held in November. Many voters changed their minds during the interim period. The 95% CI computed from data collected in September was inappropriately used to predict voting results two months later.

INTERPRETING CONFIDENCE INTERVALS IN CONTEXT

You flip a coin 20 times and observe 16 heads and 4 tails. Enter the proportion 16/20 into an appropriate program, and the 95% CI for the proportion of heads in the population will be calculated to range from 57.8% to 92.5%. (There are multiple methods for making the calculation, so your program may give a slightly different result.)

A fair coin tossed fairly will land on heads 50% of the time. But the 95% CI computed above does not include 50%. Should you therefore conclude with 95% confidence that the coin is unfair? Maybe not. It depends. . .

Accounting for prior knowledge

Your interpretation of the data depends on the context of the data collection. Consider these two scenarios:

- You've examined the coin and are 100% certain it is an ordinary coin. You also did the coin flips and recorded the results yourself, so you are sure there is no trickery. In this case, you can be virtually 100% certain that this streak of many heads is just a chance event. Regardless of the confidence interval, you'll conclude that the coin is fair. Even if you aren't 100.00000% confident in that conclusion, you'll be way more than 95% confident.
- If the coin flips were part of a magic show, you can be pretty sure that trickery invalidates the assumptions that the coin is fair, the tosses are random, and the results are accurately recorded. You'll conclude that the coin flipping was not fair.

In both cases, your interpretation of the data changes based on your prior knowledge. All good scientists informally account for prior knowledge and theory when they evaluate new data. Some also use an alternative approach to data analysis, called *Bayesian statistics*, which formally combines the current evidence or data with your prior knowledge from theory or existing data to compute a *credible interval*. Credible intervals have some similarity to confidence intervals, but there are definite distinctions which won't be explained in this book.

Accounting for multiple comparisons

Let's continue the coin-flipping example of the previous section. What if everyone in a class of 200 students had flipped the coin 20 times and the run of 16 heads was the most heads obtained by any student? You wouldn't conclude that the coin was unfair. You'd expect some students to end up with more heads and some with more tails, and you can't interpret the confidence interval at face value. More on this in Chapter 17.

CONFIDENCE INTERVALS FOR OTHER KINDS OF DATA

Both examples in this chapter (premature babies and polling voters) had outcomes with two possible values, so the data can be expressed as a proportion. But the idea of a confidence interval is far more general than that. Below are several brief examples pointing out that the concept of a confidence interval is widely used.

Confidence interval of a count

Imagine that you have counted 120 radioactive counts in one minute. Radioactive counts occur randomly and independently, and the average rate doesn't change over time (within a reasonable time frame much shorter than the isotope's half-life). We know the true average rate is unlikely to equal exactly 120 counts per minute. Statistical calculation using something called the Poisson distribution (which won't be explained in this book) lets you compute a confidence interval for the population value from that single value determined in a sample. Given a set of assumptions not detailed here (but explained in Chapter 6 of Motulsky, 2014), the 95% CI for the average number of counts per minute ranges from 99.5 to 143.5.

Confidence interval of percent survival

Figure 4.2 shows a survival curve. This book won't give any details, but this kind of graph is shown frequently in reports of clinical trials. The horizontal (X) axis shows elapsed time, and the vertical (Y) axis shows the percentage of animals or people still alive at that time. Every time someone dies, the survival percentage decreases, so you see a downward step in the graph.

Actually, the term *survival curve* is a bit limiting, because these kinds of graphs are used to plot time to any well-defined event. The event is often death, but it could be occlusion of a vascular graft, first metastasis, or rejection of a transplanted kidney. The same principles are used to plot time until a light bulb burns out, until a router needs to be rebooted, until a pipe leaks, and so on. In these cases, the term *failure time* is used instead of survival time. The event does not have to be dire, however. It could be restoration of renal function, discharge from a hospital, or graduation. But the event must be a one-time event. Special methods are needed when events can recur.

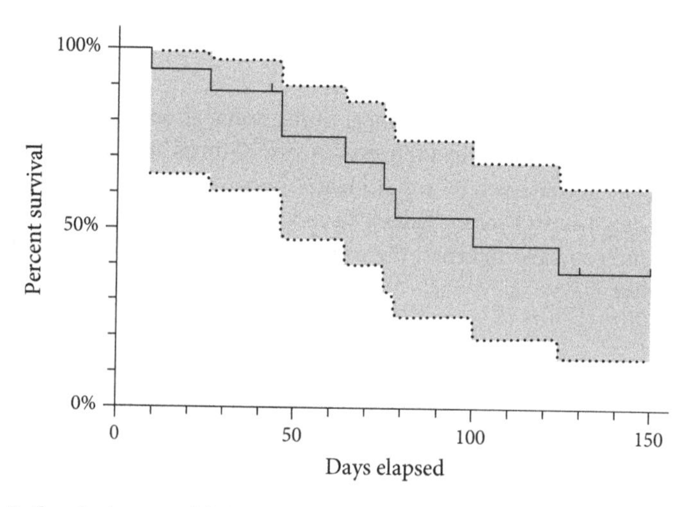

Figure 4.2. Survival curve with 95% CI.

The solid line shows the percent survival at various times. The shaded area shows the 95% CI of the percent survival.

The solid line in Figure 4.2 shows the observed survival in the sample actually studied. At time zero, we know by definition that survival equals 100%, so no confidence interval is needed there. But after that, we know that the survival of our sample is unlikely to match the average survival in the entire population. Statistical methods not explained here compute the confidence intervals shown as the shaded area in Figure 4.2. Given a set of standard assumptions, we can be 95% sure that the true average survival at each time is within the shaded area.

Confidence interval of a slope

Figure 4.3 shows data you'll see again in Chapters 22 and 23. Briefly, each circle represents data for one person. The X axis plots the lipid composition of that person's muscle (determined by a tiny biopsy). The Y axis plots that person's sensitivity to insulin. The solid line shows the line fit by linear regression (explained in Chapter 23). Its slope is 37.21 mg/m²/min per 1% increase in fatty acid. We know that the true population slope is unlikely to equal exactly this value determined from our sample of 13 people. Statistical calculations let us express the result as a confidence interval, which ranges from 16.8 to 57.7 mg/m²/min per 1% increase in fatty acid. Given a set of standard assumptions, detailed in Chapter 23, we can be 95% sure that this range includes the true population value.

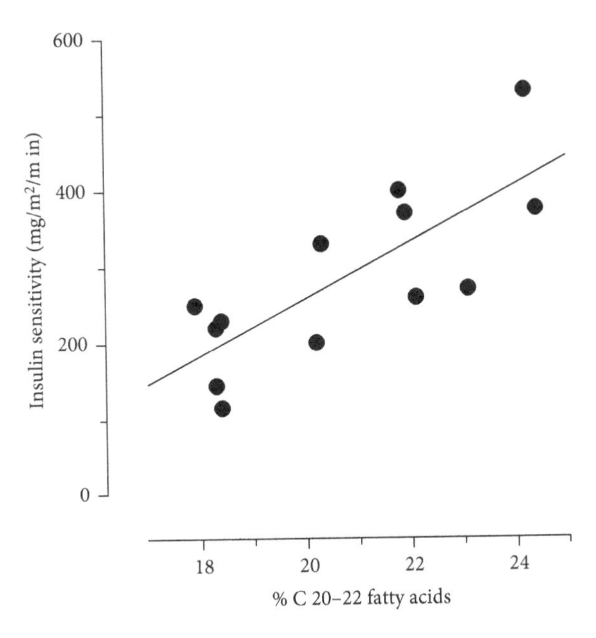

Figure 4.3. Linear regression.

Each dot shows results from one person. The solid line shows the best-fit line as determined by linear regression (see Chapter 23). The slope of this line is 37.21 mg/m²/min per 1% increase in fatty acid, and the 95% CI ranges from 16.8 to 57.7 21 mg/m²/min per 1% increase in fatty acid.

Confidence interval of a mean

If you compute the mean (average) of a set of continuous values (e.g., weights), it is possible to compute a 95% CI for the population mean. Chapter 10 will explain the details.

LINGO

Confidence intervals versus confidence limits

Each end of a confidence interval is called a *confidence limit*. The *confidence interval* is a range that extends from one confidence limit to the other.

Estimate

The sample proportion is said to be a *point estimate* of the true population proportion. The confidence interval covers a range of values and so is said to be an *interval estimate*.

Note that *estimate* has a special meaning in statistics. It does not mean an approximate calculation or an informed hunch, but rather is the result of a defined calculation. The term *estimate* is used because the value computed from your sample is only an estimate of the true value in the population (which you can't know).

Confidence level

A 95% CI has a confidence level of 95%. If you generate a 99% CI, the confidence level equals 99%. The term *confidence level* is used to describe the desired amount of confidence.

Inferential versus descriptive statistics

A confidence interval lets you make inferences about the distribution of a population by analyzing a sample. Much of this book is about this kind of *inferential statistics*. In contrast, *descriptive statistics* simply describe your sample without trying to make any inferences about the population from which the sample came.

Binomial distribution

If you flip a coin fairly, there is a 50% probability that it will land on heads and a 50% probability that it will land on tails. This means that in the long run, it will land on heads as often as it lands on tails. But in any particular series of tosses, you may not see equal numbers of heads and tails, and you may even see only heads or only tails. The *binomial distribution* calculates the likelihood of observing any particular outcome when you know its proportion in the overall population and thus answers questions like these:

- If you flip a coin fairly 10 times, what is the chance of observing exactly seven heads?

- If, on average, 5% of the patients undergoing a particular operation get an infection, what is the chance that 10 of the next 30 patients will get an infection?
- If 40% of voters are Democrats (one of the two major political parties in the United States) what is the chance that exactly 45% of a random sample of 600 US voters will be Democrats?

The *cumulative binomial distribution* answers questions like these:

- If you flip a coin fairly 10 times, what is the chance of observing seven *or more* heads?
- If, on average, 5% of the patients undergoing a particular operation get infected, what is the chance that 10 *or more* of the next 30 patients will get infected?
- If 40% of voters are Democrats, what is the chance that *at least* 45% of a random sample of 600 voters will be Democrats?

This book won't show you how to answer these questions, but you can do so with a helpful web calculator found at www.graphpad.com/quickcalcs/probability1/.

COMMON MISTAKES

Mistake: Using 100 as the denominator when the value is a percentage

The calculation of the confidence interval depends on the sample size. If the proportion is expressed as a percentage, it is easy to mistakenly enter 100 as the sample size. If you do this, you'll get the wrong answer unless the sample size actually is 100 or close to it.

Mistake: Computing binomial confidence intervals from the percent change in a continuous variable

The methods explained in this chapter apply when there are two possible outcomes (i.e., binomial outcomes), so the result is expressed as a proportion of the time that one particular outcome occurs. This value is often expressed as a percentage. But beware of percentages, which can also be used to express continuous data. Results such as percent change in weight, percent change in price, or percent change in survival time cannot be analyzed with the methods explained in this chapter. If you try to do so, the results won't make any sense.

Q & A

Which is wider, a 95% CI or a 99% CI?

To be more certain that an interval contains the true population value, you must generate a wider interval. A 99% CI is wider than a 95% CI. See Figure 4.4.

Is it possible to generate a 100% CI?

A 100% CI would have to include every possible value, so it would always extend from 0.0% to 100.0% and not be the least bit useful.

How do confidence intervals change if you increase the sample size?

The width of the confidence interval is approximately proportional to the reciprocal of the square root of the sample size. So if you increase the sample size by a factor of 4, you can expect to cut the length of the confidence interval in half. Figure 4.5 illustrates how the confidence interval gets narrower as the sample size increases.

Can you compute a confidence interval for a proportion if you know the sample proportion but not the sample size of the sample?

No. The width of the confidence interval depends on the sample size (n). See Figure 4.5.

Why isn't the confidence interval symmetrical around the observed proportion?

Because a proportion cannot go below 0.0 or above 1.0, the confidence interval will be lop-sided when the sample proportion is far from 0.50 or the sample size is small. See Figure 4.6.

You expect the population proportion to be outside your 95% CI in 5% of samples. Will you know when this happens?

No. You don't know the true value of the population proportion (except when doing simulations), so you won't know if it lies within your confidence interval or not.

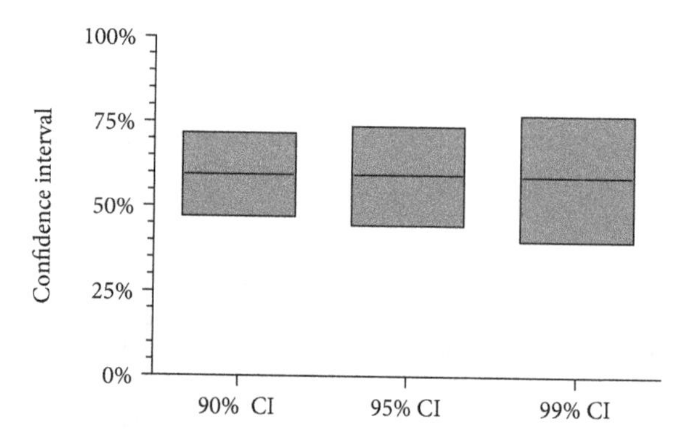

Figure 4.4. When you choose to have less confidence, the confidence interval is narrower.

Each bar represents a sample with a 60% success rate and a sample size (n) of = 40. The graph illustrates the 90%, 95%, and 99% CIs for the success rate in the population from which the data were drawn.

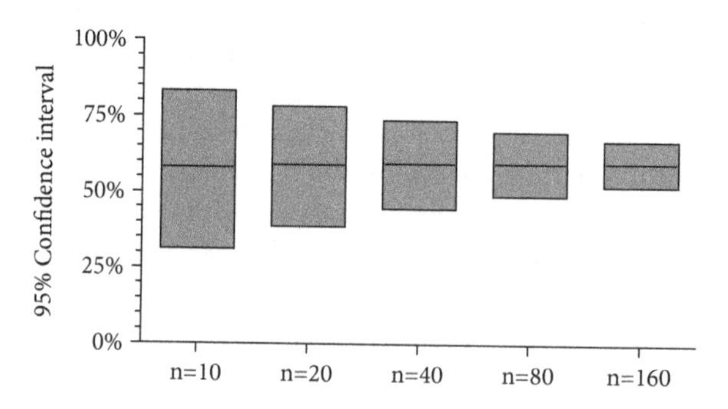

Figure 4.5. When the samples are larger, the confidence interval is narrower.

Each bar represents a sample with a 60% success rate, and the graph shows the 95% CIs for the success rate in the population from which the data were drawn.

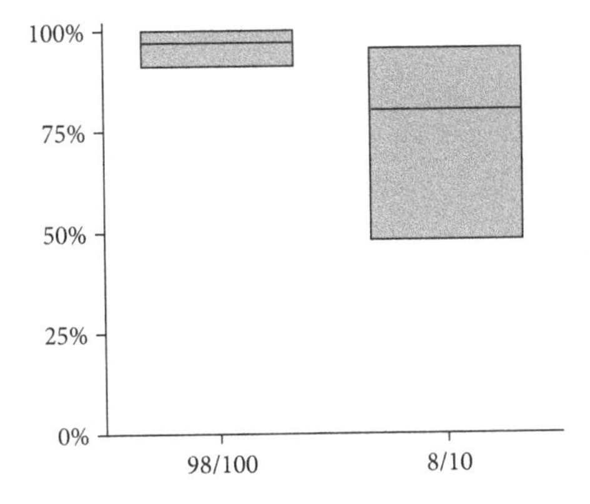

Figure 4.6. Asymmetrical confidence interval.

If the proportion is far from 50%, the 95% CI of a proportion is noticeably asymmetrical, especially when the sample size is small.

CHAPTER SUMMARY

- A fundamental concept in statistics is the use of a confidence interval to analyze a single sample of data to make conclusions about the population from which the data were sampled.
- To compute a confidence interval of a proportion, you only need the two numbers that form its numerator and its denominator.
- Given a set of assumptions, 95% of the 95% confidence intervals will include the true population value. You'll never know if a particular confidence interval is part of that 95% or not.
- There is nothing magic about 95%. Confidence intervals can be created for any degree of confidence, but 95% is used most commonly.
- The width of a confidence interval depends, in part, on the sample size. The interval is narrower when the sample size is larger.
- The concept of a confidence interval is general and can apply to almost any outcome. It is not limited to proportions.

CHAPTER 5

Types of Variables

Different kinds of data require different kinds of statistical analyses, so it helps to understand the different kinds of variables.

CONTINUOUS VARIABLES

A variable that can have an infinite number of possible values, often including zero or negative values, is called a *continuous variable*. Continuous variables can be *intervals* or *ratios*, as explained in the next sections.

Variables on an interval scale

Temperature in degrees Celsius is termed an *interval variable,* because a difference (interval) between two temperatures means the same thing no matter where you start. For example, the 10°C difference between the temperatures of 100°C and 90°C has the same meaning as the 10°C difference between the temperatures of 90°C and 80°C.

Calculating the ratio of two temperatures is not useful. The problem is that the definition of zero is arbitrary. A temperature of 0.0°C is defined as the temperature at which water freezes and certainly does not mean there is no temperature. A temperature of 0.0°F is a completely different temperature (−17.8°C). Because the zero point is arbitrary (and doesn't mean no temperature), it would make no sense to compute ratios of temperatures. A temperature of 100°C (or 212°F) is not twice as hot as a temperature of 50°C (or 122°F).

Interval variables should not be plotted on bar graphs. Figure 5.1A, for example, is misleading, because it invites you to compare the relative heights of the bars and thus think about the ratio of the values, which is not useful. Figure 5.1B uses a different baseline to demonstrate the differences. The bar for the canary is about three times as high as the bar for the platypus, but this ratio (indeed, any ratio) is not useful. Figure 5.1C illustrates the most informative way to graph these values. The use of points (rather than bars) doesn't suggest thinking in terms of a ratio. A simple table might be better than any kind of graph for these values.

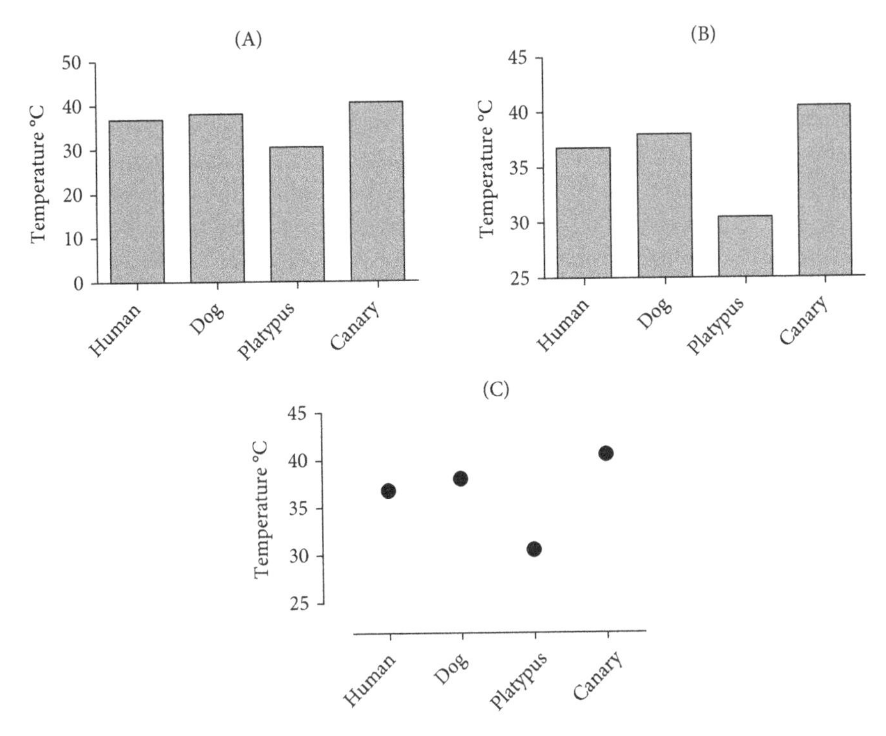

Figure 5.1. Body temperature of four species.

Graph A is misleading. It invites you to compare the relative heights of the bars. But because a temperature of 0°C does not mean "no temperature," the ratio of bar heights is not a meaningful value. Graph B uses a different baseline to emphasize the differences. The bar for the canary is about three times as high as the bar for the platypus, but this ratio (indeed, any ratio) can be misleading. Temperature in degrees Celsius is an interval variable, not a ratio variable. Graph C shows the most informative way to graph these values.

Variables on a ratio scale

With a *ratio variable*, zero is not arbitrary. Zero height really is no height. Zero weight is no weight. And zero enzyme activity is no enzyme activity. So height, weight, and enzyme activity are ratio variables.

As the name suggests, it can make sense to compute the ratio of two ratio variables. A weight of 4 grams is twice the weight of 2 grams, because weight is a ratio variable. But a temperature of 100°C is not twice as hot as a temperature of 50°C, because temperature in degrees Celsius is not a ratio variable. Note, however, that temperature in Kelvin (°K) is a ratio variable, because 0.0°K really does mean (at least to a physicist) no temperature. Temperatures in Kelvin are far removed from temperatures that we ordinarily encounter, so they are rarely used in biology.

Of course, you can also meaningfully compute the difference between ratio variables.

ORDINAL VARIABLES

An *ordinal variable* expresses rank. The order matters, but not the exact value. For example, pain is expressed on a scale of 1 to 10. A score of 7 means more pain than a score of 5, which is more pain than a score of 3. But it doesn't make sense to compute the difference between two values, because the difference between a score of 7 and a score of 5 may not be comparable to the difference between a score of 5 and a score of 3. The values simply express an order. Another example would be movie or restaurant ratings from one to five stars.

NOMINAL VARIABLES

A variable that can take on only one of a defined number of possible values is called a *categorical variable*, or a *nominal variable*. Examples include eye color (e.g., blue, green, or brown) and marital status (e.g., married, separated, divorced, widowed, or "it's complicated").

When a categorical variable has only two possible outcomes, it is called a *binary variable*, or (when encoded as 0 or 1) a *dummy variable*. Examples include alive or dead, heads or tails, and pass or fail. Even if a nominal variable is coded numerically (1 = male, 2 = female), it is still a nominal variable.

Q & A

Are the various variables types really that distinct always?

No. The categories are nowhere near as distinct as they may sound (Velleman & Wilkinson, 1993). Here are some ambiguous situations:

- Color. In a psychological study of perception, different colors would be regarded as categories, so color would be a nominal variable. But monochromatic colors can be quantified by wavelength and thus can be considered as ratio variables (although a wavelength of zero is impossible). Alternatively, you could rank the wavelengths and consider color to be an ordinal variable.
- Cells counted in a certain volume. This has all the properties of a ratio variable, except that the number must be an integer. This situation is similar to the examples outlined in Chapter 6.
- Percentages. Outcomes measured as ratio or interval variables are often transformed so that they can be expressed as percentages. For example, a pulse rate (heartbeats per minute, a ratio variable) could be normalized to a percentage of the maximum possible pulse rate. A discrete outcome with a set of mutually exclusive categories can also be expressed as a percentage or proportion—for example, the percentage of transplanted kidneys that are rejected within a year of surgery. But these two situations are very different, and different kinds of statistical analyses are needed.

What calculations are valid with which types of variables?

Table 5.1 summarizes which kinds of calculations are meaningful with which kinds of variables. It refers to the standard deviation and the coefficient of variation, both of which are explained in Chapter 7, as well as to the standard error of the mean, which is explained in Chapter 10.

OK TO COMPUTE	NOMINAL	ORDINAL	INTERVAL	RATIO
Frequency distribution	Yes	Yes	Yes	Yes
Median and percentiles	No	Yes	Yes	Yes
Add or subtract	No	No	Yes	Yes
Ratio	No	No	No	Yes
Mean, standard deviation, standard error of the mean	No	No	Yes	Yes

Table 5.1. Calculations that are meaningful with different kinds of variables.
The standard deviation and coefficient of variation will be explained in Chapter 7. The standard error of the mean will be explained in Chapter 10.

CHAPTER SUMMARY

- Ratio variables are variables for which zero is not an arbitrary value. Weight is a ratio variable, because weight = 0 means there is no weight. Temperature in degrees Celsius (or degrees Fahrenheit) is not a ratio variable, because 0°C (or 0°F) does not mean there is no temperature.
- Interval variables are variables for which a certain difference (interval) between two values is interpreted identically no matter where you start but for which zero is defined arbitrarily.
- It does not make any sense to compute a coefficient of variation or to compute ratios of interval variables. Nor does it make sense to plot this kind of data on bar graphs.
- The distinctions among the different kinds of variables are not quite as crisp as they sound.

CHAPTER 6

Graphing Variability

GRAPHING DATA TO SHOW SCATTER OR DISTRIBUTION

Sample data

Mackowiak, Wasserman, and Levine (1992) measured body temperature from healthy individuals to see what the normal temperature range really is. A subset of 12 of these values is shown in Table 6.1. These values—and the suggestion to use them to explain basic statistical principles—come (with permission) from Schoemaker (1996).

Scatter plots

Figure 6.1 plots the temperature data as a *column scatter plot*. The left side of the graph shows all 130 values. The right side of the graph shows the randomly selected subset of 12 values (Table 6.1), which we will analyze separately. Each value is plotted as a symbol. Within each half of the graph, symbols are moved to the right or the left to prevent too much overlap. The horizontal position is arbitrary, but the vertical position, of course, denotes the measured value.

This kind of column scatter plot, also known as a *dot plot*, demonstrates exactly how the data are distributed. You can see the lowest and the highest values as well as the distribution. A horizontal line is sometimes drawn at the *mean* (average) or *median* (50th percentile).

With huge numbers of values, column scatter plots get unwieldy because of so many overlapping points. The left side of Figure 6.1, with 130 circles, is pushing the limits for this kind of graph.

Box-and-whiskers plots

A *box-and-whiskers plot* gives you a good sense of the distribution of data without showing every value (see Figure 6.2). Box-and-whiskers plots work great when you have too many data points to show clearly on a column scatter plot but don't want to take the space to show a full frequency distribution.

37.0
36.0
37.1
37.1
36.2
37.3
36.8
37.0
36.3
36.9
36.7
36.8

Table 6.1. The body temperature of 12 individuals in degrees Celsius.

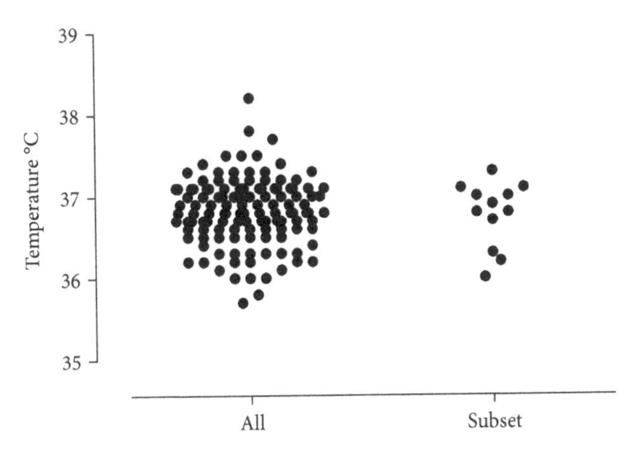

Figure 6.1. Column scatter plot of body temperatures.
(Left) The entire data set, with n = 130. (Right) A randomly selected subset, with n = 12. In a column scatter graph, the vertical position of each symbol denotes its value. The horizontal position (within each lane) is adjusted (jittered) to prevent points from overlapping (too much).

A horizontal line marks the median (50th percentile) of each group. The boxes extend from the 25th to the 75th percentiles and therefore contain half the values. A quarter (25%) of the values are higher than the top of the box, and a quarter (25%) of the values are below the bottom of the box. If you are not already familiar with percentiles, they will be discussed in more detail in the next chapter.

The whiskers can be graphed in various ways. The box-and-whiskers plot on the left in Figure 6.2 plots the whiskers down to the 5th and up to the 95th percentiles and individual dots for values lower than the 5th and higher than the 95th percentiles. The box-and-whiskers plot in the middle in Figure 6.2 plots the

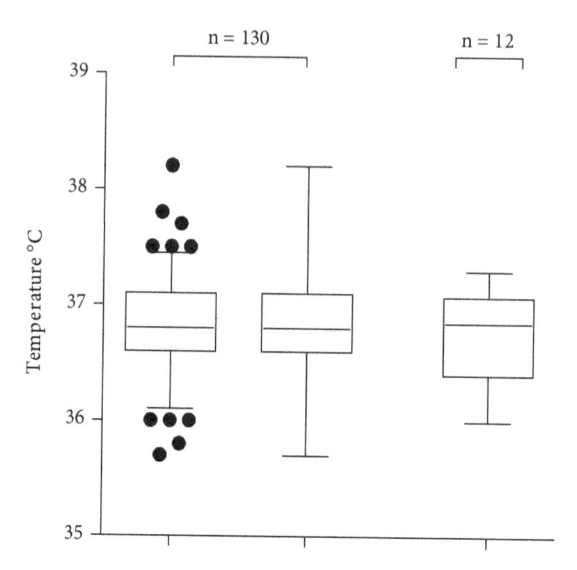

Figure 6.2. Box-and-whiskers plots.

(Left) Box-and-whiskers plot of the entire data set. The whiskers extend down to the 5th percentile and up to the 95th percentile, with individual values showing beyond that. (Middle) Box-and-whiskers plot showing the range of all the data. (Right) Box-and-whiskers plot for a subset of 12 values.

whiskers down to the smallest value and up to the largest, so it doesn't plot any individual points. Whiskers can be defined in other ways as well.

Frequency distribution histograms

A *frequency distribution histogram* lets you see the distribution of many values. Divide the range of values into a set of smaller ranges (bins), and then graph the number of values (or the fraction of values) in each bin. Figure 6.3 displays frequency distribution histograms for the temperature data. If you add the height of all the bars, you'll get the total number of values. If the graph plots fractions or percentages instead of the number of values, then the sum of the height of all bars will equal 1.0, or 100%.

The trick in constructing frequency distributions is deciding how wide to make each bin. The three graphs in Figure 6.3 use different bin widths. The graph on the top left has too few bins (each bin covers too wide a range of values), so it doesn't really show you enough detail about how the data are distributed. The graph on the bottom has too many bins (each bin covers too narrow a range of values), so it shows you too much detail (in my opinion). The graph on the top right seems the most useful.

Watch out for the term *histogram*. It is usually defined as a frequency distribution plotted as a bar graph, as illustrated in Figure 6.3. Sometimes, however, the term is used more generally to refer to any bar graph, even one that is not a frequency distribution.

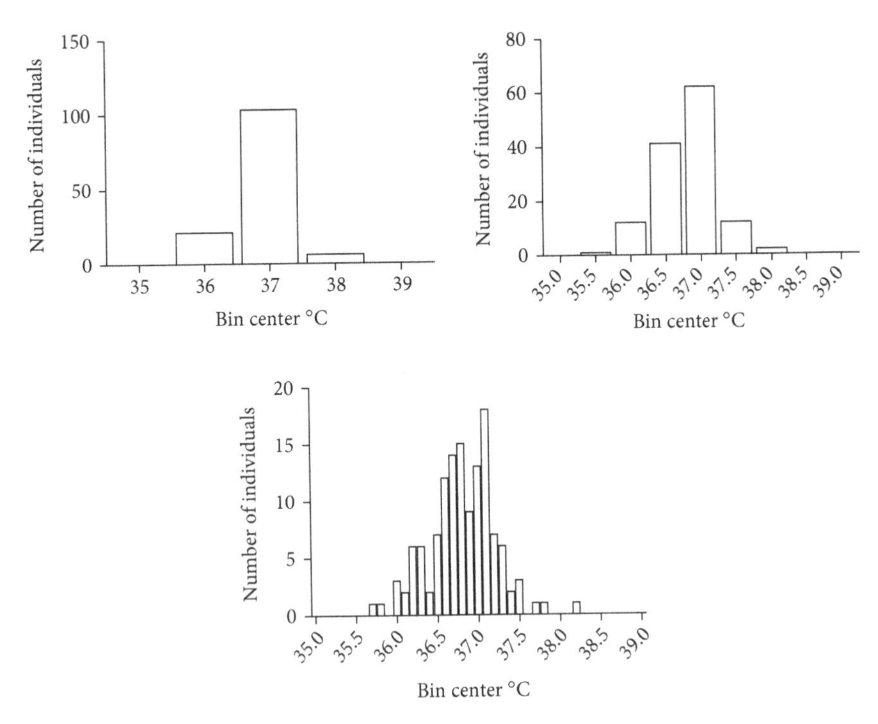

Figure 6.3. Frequency distribution histograms of the temperature data with various bin widths.

Each bar plots the number of individuals whose body temperature is in a defined range (bin). With too few bins (top left), you don't get a sense of how the values vary. With too many bins (bottom), the graph shows too much detail for most purposes. Compare those two graphs with the one in which the number of bins that seems more appropriate (top right). The centers of each range, the bin centers, are labeled.

WATCH OUT FOR PREPROCESSED DATA

Published graphs don't always plot the data that were actually collected and instead sometimes plot the result of computations. These calculations can be useful, but they can also be misleading.

Beware of filtering out impossible values

Data sets are often screened to remove impossible values. Weights can't be negative. The year of death cannot be before the year of birth. The age of a child cannot be greater than the age of the mother. Men can't be pregnant. It makes no sense to run statistical analyses on obviously incorrect data. These values must be fixed (if the mistake can be traced) or removed and documented.

But beware! Eliminating "impossible" values can also prevent you from seeing important findings. In 1985, researchers first noticed the drop in ozone levels over Antarctica (which turned out to be caused by chlorofluorocarbons). But measurements from a satellite demonstrated no such drop. Why the discrepancy?

One story is that the satellite had been reporting a drop in ozone levels for years, but these data were automatically excluded from the analyses because they were considered to be impossible values. This story has been disputed (Pukelsheim, 1990), but it is illuminating even if false. One must be very careful when automatically filtering out "impossible" values.

Beware of adjusting data

The values analyzed by statistical tests are often not direct experimental measurements. It is common to adjust the data, so it is important to think about whether these adjustments are correct and whether they introduce errors. If the numbers going into a statistical test are questionable, then so are the statistical conclusions.

In some cases, many adjustments are needed, and these adjustments can have a huge impact. As a case in point, NASA provides historical records of temperature. An examination of how temperature has changed over the past century has convinced most scientists that the world is getting warmer. The temperature record, however, requires many corrections to account for the fact that cities are warmer than surrounding areas (because of heat from buildings), that different kinds of thermometers have been used at different historical times, that the time of day when the temperatures have been taken is not consistent, and so on (Goddard, 2008). The net result of all these adjustments pushed earlier temperatures down by almost 1°C, which is about half of the observed temperature increase in the past century. These adjustments require judgment, and different scientists may do the adjustments differently. And anyone interpreting the data must understand the contribution of these adjustments to the overall observed effect as well as the degree to which these adjustments could possibly be biased by the scientists' desire to make the data come out a certain way.

When interpreting published data, ask about how the data were adjusted before they were graphed or entered into a statistics program.

Beware of smoothing

When plotting data that change over time, it is tempting to remove much of the variability in order to make the overall trend more visible. This can be accomplished by plotting a *rolling average*, also called a *moving average* or *smoothed data*. For example, each point on the graph can be replaced by the average of it and the three nearest neighbors on each side. The number of points averaged to smooth the data can range from two to many. If more points are included in the rolling average (e.g., 10 on each side), the curve will be smoother. Smoothing methods differ in how neighboring points are weighted.

Smoothed data should never be entered into statistical calculations. If you enter smoothed data into statistics programs, the results reported by most tests will be invalid. Smoothing removes information, so most analyses of smoothed data cannot be interpreted at face value.

Beware of variables that are the ratio of two measurements

Often, the value that you care most about is the ratio of two values. For example, divide enzyme activity or the number of binding sites by the cell count or the protein concentration. Calculating the ratio is necessary to express the variable in a way that can be interpreted and compared—for example, as enzyme activity per milligram of protein. The numerator is usually what you are thinking about and what the experimenter spent a lot of time and effort to measure. The denominator can seem like a housekeeping detail. But of course, the accuracy of the ratio depends on the accuracy of both the numerator and the denominator.

Beware of normalized data

Some scientists transform (normalize) data so that all values are between 0% and 100%. When you see these kinds of data, you should wonder about how the values defining 0% and 100% were chosen. Ideally, the definitions of 0% and 100% are based on theory or on control experiments with plenty of precise replicates. If 0% and 100% are not clearly defined, or if they seem to be defined sloppily, then the normalized values won't be very useful. Ask how many replicates there were of the control measurements that define 0% and 100%.

LINGO

Error

In Table 6.1, the temperatures range from 36.0°C to 37.3°C. There are three possible reasons for this variation;

- Biological variability. People (and animals, and even cells) are different from one another, and these differences are important! Moreover, people (and animals, and again, even cells) vary over time and when exposed to different environments. In biological and clinical studies, much or even most of the scatter is often caused by biological variation.
- Imprecision, or *experimental error*. Reading older thermometers takes a bit of judgment, and the process can be prone to error. Some statistics books (especially those designed for engineers) implicitly assume that most variability is the result of imprecision. In medical studies, biological variation often contributes more variation than experimental imprecision does.
- Mistakes and glitches. Maybe a value was written down incorrectly or the thermometer wasn't positioned properly.

The word *error* is often used to refer to all three of these sources of variability. Note that this statistical use of the term is quite different from the everyday use of the word (to mean a mistake). The terms *scatter* and *variability* are more understandable than the term *error,* but statistics books tend to use *error* when referring to any kind of variation not explained by treatment effects.

Bias

As used in statistics, the word *bias* refers to anything that leads to systematic errors, not just the preconceived notions of the experimenter. Bias can be caused by any factor that consistently alters the results: the proverbial thumb on the scale, defective thermometers, bugs in computer programs (maybe the temperature was measured in Celsius and the program that converted the values to

Fahrenheit was buggy), the placebo effect, and so on. Data collected using biased methods are usually not accurate.

Precision

The term *precision* refers to how reproducible a method is, not to how accurate it is. Compare repeated measurements to see how precise (or repeatable, or reproducible) the measurements are. A method is precise when repeated measurements give very similar results. Note, however, that a set of measurements could be quite precise without being accurate in the presence of bias.

COMMON MISTAKES

Mistake: Expressing variability by showing or graphing the mean and an error bar without considering showing the actual data

Error bars will be explained in Chapter 11. Plotting the mean (or median) and an error bar can be a concise way to show data but tends to show too little detail, so this can be misleading.

Mistake: Graphing data that has been preprocessed without explaining what you did

In some situations, it makes sense to adjust or normalize data. In some cases, it may even make sense to smooth data. But it is essential that figures be properly labeled so that people will know exactly what is being graphed.

Mistake: Using the term *error* without explaining exactly what you mean

The term *error* has many meanings and is best avoided. When you see this word, always make sure you understand which meaning is being used.

Mistake: Mixing up the terms *bias*, *inaccurate*, and *imprecise*

The terms *bias*, *inaccurate*, and *imprecise* have distinct meanings but are often used sloppily.

Q & A

What determines the horizontal position of points in a column scatter plot (dot plot)?

The symbols are moved horizontally to avoid overlap, so you can see all the points. There is no other reason for, or meaning of, the horizontal position.

Can a measurement from a biased method be accurate?

Sure. If a method is biased, this means that on average the measurements it produces will be wrong. But in some cases, bias in one direction may balance experimental error in the other direction, resulting in a very accurate measurement.

CHAPTER SUMMARY

- Many scientific variables are continuous.
- When presenting continuous data, consider creating a graph that shows the scatter of the data.
- A box-and-whiskers plot is useful when there are too many points to plot every value.
- A frequency distribution histogram shows the distribution of values in a large data set.
- It is often useful to filter, adjust, smooth, or normalize the data before further graphing and analysis. These methods can be abused, however. Think carefully about whether these methods are being used effectively and honestly.
- The term *error* has many meanings. Avoid using it. When others use this term, make sure you understand what they mean.
- The term *bias* refers to any factor that leads to systematic errors.
- The term *precision* refers to how reproducible a method is, not to how accurate it is.

CHAPTER 7

Quantifying Variation

RANGE

In Table 6.1, the lowest value is 36.0°C, and the largest value is 37.3°C. In plain language, one might say the temperatures range from 36.0°C to 37.3°C. But in statistics, the term *range* usually refers to difference between the largest and the smallest values in a sample. For Table 6.1, the range equals 37.3°C minus 36.0°C, which is 1.3°C. Because both the smallest and the largest values are expressed in the same units as the data, the range is also expressed in those units.

PERCENTILES

You probably are already familiar with *percentiles*. For example, the 95th percentile is the value that is larger than 95% of the values in the data set and smaller than 5% of the values.

The *median* is the 50th percentile. Rank the values from lowest to highest, and the median is the middle value. If the number of values is even, the median is defined as the average of the two middle values. For the n = 130 data set from Chapter 6, the median is the average of the 65th and the 66th ranked values, or 36.85°C. Half the values are larger than the median, and half the values are lower.

How are the percentiles actually calculated? It is trickier than you might guess, with eight different methods to choose from (Harter, 1984). All methods compute the same result for the 50th percentile (the median) but not necessarily for other percentiles, especially with small samples. With large data sets, however, the results will be similar.

INTERQUARTILE RANGE

The 25th and 75th percentile values are called *quartiles*. The *interquartile range* is defined by subtracting the 25th percentile of a set of numbers from the 75th percentile of those values. Because both quartiles (25th and 75th percentiles) are expressed in the same units as the data, the interquartile range is also expressed in the same units.

For the body temperature data (n = 12 subset), the 25th percentile is 36.4°C, and the 75th percentile is 37.1°C. Therefore, the interquartile range is 0.7°C. For the full data set (n = 130), the 25th percentile is 36.6°C, and the 75th percentile is 37.1°C. So the interquartile range is 0.5°C.

FIVE-NUMBER SUMMARY

The distribution of a set of numbers is sometimes summarized with five values, known as the *five-number summary*: the minimum, the 25th percentile, the median, the 75th percentile, and the maximum.

STANDARD DEVIATION

The variation of values around their *mean* (average) can be quantified as the *standard deviation,* which is expressed in the same units as the data. *In this book, I use the* abbreviation SD, but many books use the mathematical symbol *s* instead.

Interpreting the standard deviation

You can interpret the standard deviation using the following rule of thumb: About two-thirds of the observations in a population usually lie within the range defined by the mean minus 1 SD to the mean plus 1 SD. This definition is unsatisfying, however, because the word *usually is* so vague. Chapter 8 will give a more rigorous interpretation of the standard deviation for data sampled from a Gaussian distribution.

Let's turn to the larger (n = 130) sample from Chapter 6. The mean temperature is 36.82°C, and the standard deviation is 0.41°C. The middle graph of Figure 7.1 plots the mean with error bars extending 1 SD in each direction (error bars can be defined in other ways, as you'll see in Chapter 11).

The range 36.4°C to 37.2°C extends below and above the mean by 1 SD. If you compare the left and middle plots of Figure 7.1, that range includes about two-thirds of the values.

Calculate a standard deviation

You'll use a computer to calculate the standard deviation, but read on if you are curious to know where it comes from. The standard deviation is the square root of the average of the squared deviations. Huh? Follow these steps:

1. Calculate the mean (average). For the n = 12 body temperature sample in Table 6.1, the mean is 36.77°C.
2. Calculate the difference between each value and the mean.
3. Square each of those differences.
4. Add up those squared differences. For the example data, the sum is 1.767.
5. Divide that sum by n − 1, where n is the number of values. For the example, n = 12, and 1.767/11 = 0.161. This value is called the variance.

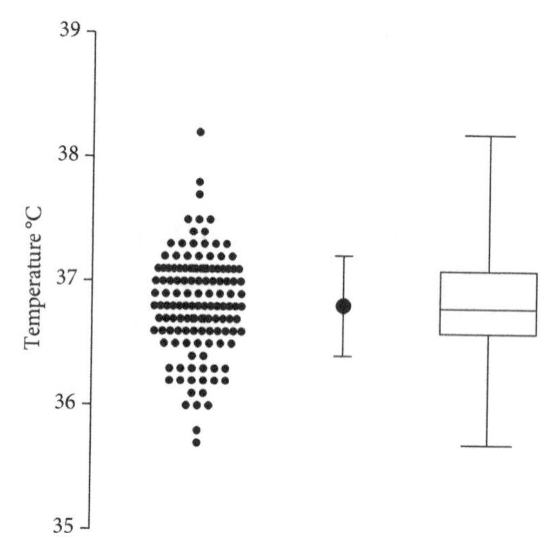

Figure 7.1. Plots of body temperature data from Chapter 6.
(Left) Individual values. (Center) The mean and standard deviation. (Right) Box-and-whiskers plot showing the five-number summary. The whiskers go down to the minimum value and up to the maximum value. The box shows the 25th and 75th percentile, and the horizontal line marks the median.

6. Take the square root of the value you calculated in step 5. The result is the standard deviation. For this example, SD = 0.40°C.

These steps can be shown as an equation, where Y_i stands for one of the n values, and $\bar{Y}$ is the mean:

$$SD = \sqrt{\frac{\sum (Y_i - \bar{Y})^2}{n-1}}$$

Why n − 1? Why not n?

When using some calculators and programs, you'll see that you have two choices for calculating the standard deviation:

- The *sample standard deviation* is the best possible estimate for the standard deviation of the entire population, as determined from one particular sample. This uses a denominator of (n − 1), where n is the sample size, so you'll sometimes see "n − 1" on the calculator or program.
- The *population standard deviation* (which uses a denominator of n) is correct for those particular values, but it cannot be used to make general conclusions from the data. One example would be when a teacher wants to quantify variation among exam scores.

Most often, a scientist's goal is to compute the standard deviation from a sample of data and make inferences about the population from which the data were sampled. So scientists almost always use the sample standard deviation computed with a denominator of (n − 1).

Standard deviation and sample size

For the full body temperature data set (n = 130), as shown in Figure 6.1, SD = 0.41°C. For the smaller sample (n = 12, randomly sampled from the larger sample), SD = 0.40°C. Many find it surprising that the standard deviation is so similar in samples of such different sizes. But in fact, this is to be expected. The standard deviation quantifies variability. As you collect larger samples, you'll be able to quantify the variability more precisely, but collecting more data doesn't change the variability among the values. As you increase the sample size, you don't expect the standard deviation to change much and are equally likely to see the standard deviation getting much larger or much smaller.

COEFFICIENT OF VARIATION

For ratio variables, variability can be quantified as the *coefficient of variation* (abbreviated CV), which equals the standard deviation divided by the mean. If the coefficient of variation equals 0.25, you know that the standard deviation is 25% of the mean.

Because the standard deviation and the mean are both expressed in the same units, the coefficient of variation is a fraction with no units. Often scientists multiply the coefficient of variation by 100, to express it as a percentage. For the preceding temperature example, the coefficient of variation would be completely meaningless. Temperature is an interval variable, not a ratio variable, because zero is defined arbitrarily (see Chapter 5). A coefficient of variation computed from temperatures measured in degrees Celsius would not be the same as a coefficient of variation computed from temperatures measured in degrees Fahrenheit. Neither coefficient of variation would be meaningful, because the idea of dividing a measure of scatter by the mean only makes sense with ratio variables, for which zero really means zero.

The coefficient of variation is useful for comparing scatter of variables measured in different units. You could ask, for example, whether the variation in pulse rate is greater than or less than the variation in the concentration of serum sodium. The pulse rate and sodium are measured in completely different units, so comparing their standard deviations would make no sense. Comparing their coefficients of variation might be useful to someone studying physiology.

LINGO

This chapter explains various ways to quantify the variation among values. But don't mix up the word *variation* (which is fairly general) and the word *variance* (which has a very specific meaning). The *variance* equals the standard deviation squared and so is expressed in the square of the units used for the data. In the example, the variance is 0.4°C × 0.4°C, which is 0.16 degrees squared. Statistical theory is based on variances rather than standard deviations, so mathematical statisticians routinely think about variances. Scientists analyzing data can usually focus on the standard deviation (or coefficient of variation) and not think about variances (or, for this example, degrees squared).

COMMON MISTAKES

Mistake: Focusing on only the mean and ignoring variability

How many breasts and testicles does the average person have? One of each! But this doesn't tell you very much. You need to think about variability, not just the averages.

Mistake: Expressing variability by showing or graphing the mean and the standard error of the mean

As will be explained in Chapter 10, the standard error of the mean does not directly quantify variability among values.

Mistake: Calculating the coefficient of variation of a variable that is not a ratio variable

With a ratio variable (defined in Chapter 5), zero means zero. Zero length means no length. Zero grams means no weight. But 0°C does not mean no temperature, so temperature in Celsius is not a ratio variable. If you calculated the coefficient of variation of temperature, the value would be meaningless.

Q & A

Can the standard deviation ever equal zero? Can it be negative?

The standard deviation will equal zero if all the values are identical, but it can never be a negative number. The standard deviation is a measure of variability, and it would be meaningless to say the variability is a negative value.

In what units do you express the standard deviation?

The standard deviation is expressed in the same units as the data.

Can the standard deviation be computed when n = 1? When n = 2?

The standard deviation quantifies variability, so it cannot be computed from a single value. However, it can be computed from two values (n = 2).

Is the standard deviation the same as the standard error of the mean?

No. They are very different. See Chapter 10.

Is the standard deviation larger or smaller than the coefficient of variation?

The standard deviation is expressed in the same units as the data. The coefficient of variation is a unitless ratio, often expressed as a percentage. Because the two are not expressed in the same units, it makes no sense to ask which is larger.

Is the standard deviation larger or smaller than the variance?

The variance is the standard deviation squared, so it is expressed in different units. It makes no sense to ask which is larger.

Will all programs compute standard deviation the same way?

The only ambiguity is whether to use n or (n − 1) in the denominator.

CHAPTER SUMMARY

- One way to summarize a set of values is to calculate the mean or median.
- The most common way to quantify scatter is with a standard deviation.
- Scientists almost always use the sample standard deviation, which has a denominator of (n − 1), rather than the population standard deviation, which has a denominator of n. The sample standard deviation can be used to make general conclusions about the population from a limited sample.
- A useful rule of thumb is that about two-thirds of the observations in a population often lie within the range defined by the mean minus 1 SD to the mean plus 1 SD.
- The variance is defined to equal the standard deviation squared. While much of the mathematics underlying statistics is developed in terms of variances, it is much easier for most people to understand data by thinking about the standard deviation and avoiding variances.
- The coefficient of variation equals the standard deviation divided by the mean. It is meaningful only for ratio variables (defined in Chapter 5).
- While it is useful to quantify variation, it is often easiest to understand the variability in a data set by seeing a graph of every data point or of the frequency distribution.

CHAPTER 8

The Gaussian Distribution

HOW THE GAUSSIAN DISTRIBUTION ARISES

You probably are already familiar with the bell-shaped Gaussian distribution, which is the basis for much of statistics.

The Gaussian distribution arises when variability is caused by the sum of many random and independent factors. Some factors will push the value higher, and others will pull it lower. Usually, the effects partly cancel out one another, so many values end up near the center (the mean). Sometimes, many random factors will tend to work in the same direction, pushing a value away from the mean. Only rarely do the majority of random factors all work in the same direction, pushing that value far from the mean. So you end up with a symmetrical distribution of values, with many values near the mean, some values farther from the mean, and very few quite far from the mean.

Variation among values will approximate a Gaussian distribution when there are many sources of variation, so long as the various contributors to that variation are added up to get the final result and the sample size is large. As you have more and more sources of scatter, the predicted result approaches a Gaussian distribution. For example, in a laboratory experiment, variation between experiments might be caused by several factors: imprecise weighing of reagents, imprecise pipetting, the random nature of radioactive decay, nonhomogeneous suspensions of cells or membranes, and so on. Variation in a clinical value might be caused by many genetic and environmental factors.

THE MEANING OF STANDARD DEVIATION IN A GAUSSIAN DISTRIBUTION

Figure 8.1 illustrates the symmetrical, bell-shaped *Gaussian distribution*. The horizontal axis shows various values that can be observed, and the vertical axis quantifies their relative frequency. The mean, of course, is the center of the Gaussian distribution. As you move away from the mean, the distribution follows its characteristic bell shape. The distribution is symmetrical, so the median and the mean are identical.

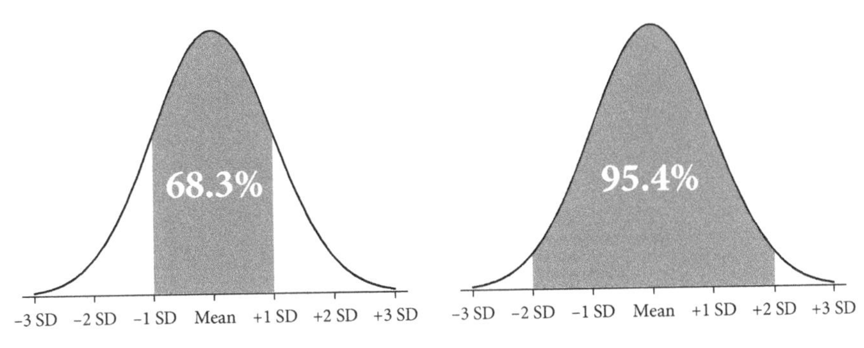

Figure 8.1. Ideal Gaussian distributions.
The horizontal axis plots various values, and the vertical axis plots their relative abundance. The area under the curve represents all values in the population. The fraction of that area within a range of values tells you how common those values are. (Left) A bit more than two-thirds of the values are within 1 SD of the mean. (Right) Slightly more than 95% of the values are within 2 SD of the mean.

The *standard deviation* (SD) is a measure of the spread or width of the distribution. The area under the entire curve represents the entire population. The graph on the left side of Figure 8.1 shades the area under the curve within 1 SD of the mean. The shaded portion is about two-thirds (68.3%) of the entire area, demonstrating that about two-thirds of the values in a Gaussian population are within 1 SD of the mean. The graph on the right side of Figure 8.1 demonstrates that about 95% of the values in a Gaussian population are within 2 SD of the mean (the actual multiplier is 1.96).

Note also that the standard deviation is the distance from the mean to the inflection point of the Gaussian distribution.

Scientific papers and presentations often show the mean and standard deviation but not the actual data. If you assume that the distribution is approximately Gaussian, you can re-create the distribution in your head. Going back to the sample body temperature data in Figure 7.1, what could you infer if you knew only that the mean is 36.8°C and the standard deviation is 0.4°C ($n = 130$)? If you assume a Gaussian distribution, you could infer that about two-thirds of the values lie between 36.4°C and 37.2°C (the mean ± 1 SD) and that 95% of the values lie between 36.0°C and 37.6°C (the mean ± 2 SD). If you look back at Figure 7.1, you can see that these estimates are fairly accurate.

WHAT A SAMPLE DRAWN FROM A GAUSSIAN DISTRIBUTION REALLY LOOKS LIKE

The ideal Gaussian frequency distribution was shown in Figure 8.1. With huge data sets, this is what you expect a sample of data drawn from a Gaussian distribution to look like. But what about smaller data sets?

The bell-shaped Gaussian distribution is an ideal distribution of the population. Unless the samples are huge, actual frequency distributions tend to be less symmetrical and somewhat jagged, as shown in Figure 8.2. Each of the eight

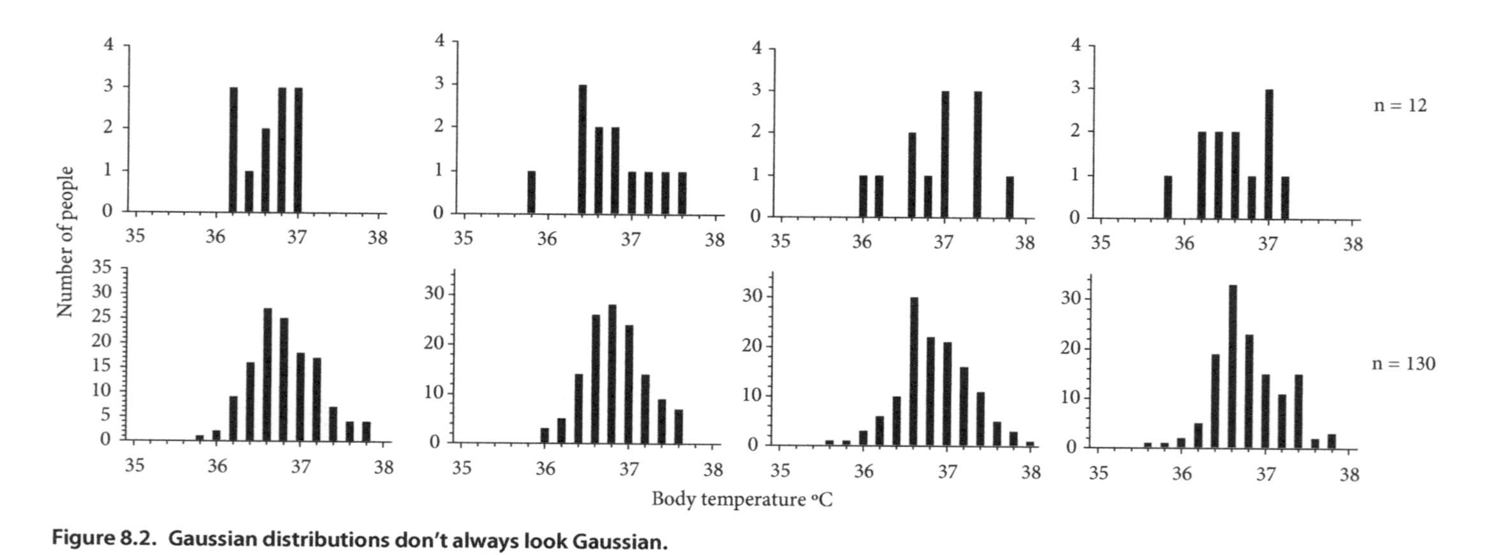

Figure 8.2. Gaussian distributions don't always look Gaussian.

All graphs show frequency distributions of values randomly chosen from a Gaussian distribution. The differences among the graphs are simply the result of random variation when sampling from a Gaussian distribution.

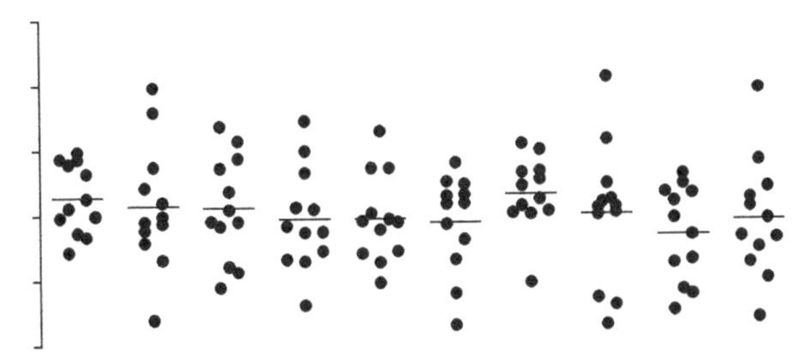

Figure 8.3. Samples from Gaussian distributions don't always look Gaussian.

All 10 samples were randomly sampled from Gaussian distributions. It is too easy to be fooled by random variation and think that the data are far from Gaussian.

frequency distributions in Figure 8.2 shows the distribution of values randomly chosen from a Gaussian distribution. Because of random sampling variation, none of these frequency distributions really looks completely bell shaped and symmetrical. Figure 8.3 makes the same point by showing individual values sampled from a Gaussian distribution.

WHY THE GAUSSIAN DISTRIBUTION IS SO CENTRAL TO STATISTICAL THEORY

The Gaussian distribution plays a central role in statistics because of a mathematical relationship known as the *central limit theorem*. You'll need to read a more theoretical book to really understand this theorem, but the following imaginary experiment gives you the basics:

1. Create a population with a known distribution that is not Gaussian.
2. Randomly pick many samples of equal size from that population.
3. Tabulate the means of these samples.
4. Graph the frequency distribution of those means.

The central limit theorem says that if your samples are large enough, the distribution of the means will approximate a Gaussian distribution even if the population is not Gaussian. Because most statistical tests (such as the t test and analysis of variance) are concerned only with differences between means (taking into account variability and sample size), the central limit theorem explains why these tests work well even when the populations are not Gaussian. How large is large enough? There is no simple answer. It depends on how asymmetrical (skewed) the distribution is.

LINGO

The Gaussian distribution is also called a *normal distribution*. The two terms are synonyms. Note that the use of the word *normal* to describe a Gaussian distribution is completely distinct from the use of that word to describe something that is healthy or common.

When the mean equals 0 and the standard deviation equals 1.0, the Gaussian distribution is called a *standard normal distribution*. Figure 8.1 would be a standard normal distribution if the labels went from −3 to +3 without using the labels SD and mean.

All Gaussian distributions can be converted to a standard normal distribution. To do this, subtract the mean from each value, and divide the difference by the standard deviation. You'll often see this value called *z*, which is the number of standard deviations a value is away from the mean. When $z = 1$, a value is 1 SD above the mean. When $z = -2$, a value is 2 SD below the mean.

COMMON MISTAKES

Mistake: Misunderstanding the word *normal*

The Gaussian distribution is also called a normal distribution. But this use of the word *normal* is very different than the usual meanings of common or disease free.

Mistake: Believing that all distributions are supposed to be Gaussian

The Gaussian distribution has mathematical properties that power many statistical analyses. But that doesn't mean that every distribution is Gaussian. Few, if any, distributions are exactly Gaussian, because that would imply some possibility of very large numbers and very negative values. But many biological distributions are approximately Gaussian.

Q & A

Is the Gaussian distribution the same as a normal distribution?

Yes, the two terms are used interchangeably. But note that the term *normal* has many other meanings as well.

Are all bell-shaped distributions Gaussian?

As you can see in Figure 8.1, the Gaussian distribution is bell shaped. But not all bell-shaped distributions are Gaussian.

Will numerous sources of scatter always create a Gaussian distribution?

No. A Gaussian distribution is formed only when each source of variability is independent and additive with the others and no one source dominates. Chapter 9 discusses what happens when the sources of variation multiply.

CHAPTER SUMMARY

- The bell-shaped Gaussian distribution is the basis for much of statistics. It arises when many random factors create variability.
- With a Gaussian distribution, about two-thirds of the values are within 1 SD of the mean, and about 95% of the values are within 2 SD of the mean.
- The Gaussian distribution is also called a normal distribution. But this use of the term *normal* is very different than the usual use of that word to mean ordinary or without disease.

- The central limit theorem explains why Gaussian distributions are fundamental to much of statistics. Basically, this theorem says that the distribution of many sample means will tend to be Gaussian even if the data are not sampled from a Gaussian distribution.
- Data sampled from Gaussian distributions don't look as Gaussian as many expect. Random sampling introduces more randomness than many appreciate.

CHAPTER 9

The Lognormal Distribution and Geometric Mean

OVERVIEW

Lognormal distributions occur very commonly in many fields of science and so are a core concept of biostatistics. If you don't recognize when data are lognormal, you are likely to misinterpret statistical results.

EXAMPLE: RELAXING BLADDERS

And you thought statistics was never relaxing!

Frazier, Schneider, and Michel (2006) measured the ability of isoprenaline (a drug that acts much like the neurotransmitter norepinephrine, also called noradrenaline) to relax the bladder muscle. The results are expressed as the EC50, which is the concentration required to relax the bladder halfway between its minimum and maximum possible relaxation. The graph on the left side of Figure 9.1 illustrates the data plotted on a linear scale. The distribution is far from symmetrical and is quite skewed. One value is far from the rest and almost looks like a mistake.

A REVIEW OF LOGARITHMS

As the name suggests, lognormal distributions are based on logarithms, so before explaining lognormal distributions, let's review logarithms.

What are logarithms?

The best way to understand logarithms is through an example. If you take 10 to the third power ($10 \times 10 \times 10$), the result is 1,000. The *logarithm* is the inverse of that power function. The logarithm (base 10) of 1,000 is the power of 10 that gives the answer 1,000. So the logarithm of 1,000 is 3. If you multiply 10 by itself three times, you get 1,000.

You can take 10 to a negative power. For example, taking 10 to the -3 power is the same as taking the reciprocal of 10^3. So 10^{-3} equals $1/10^3$, or 0.001. The logarithm of 0.001 is the power of 10 that equals 0.001, which is -3.

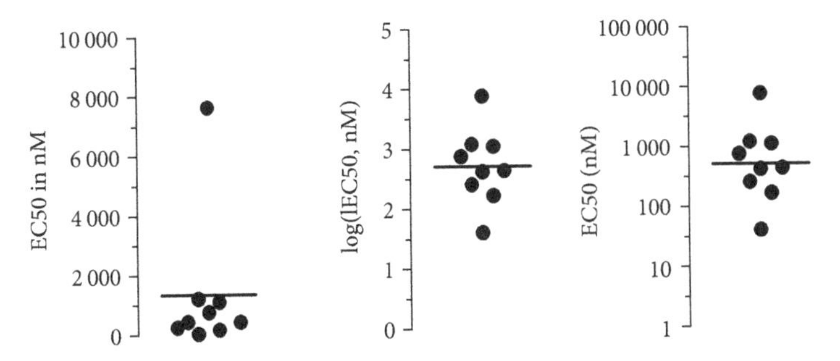

Figure 9.1. Lognormal data.

These data demonstrate the EC50 of isoprenaline for relaxing bladder muscles (Frazier, Schneider, & Michel, 2006). Each dot indicates data from the bladder of a different animal. The EC50 is the concentration required to relax the bladder halfway between its minimum and maximum possible relaxation, measured in nanomoles per liter (nM). (Left) Plot of the original concentration scale. The data are far from symmetrical, and the highest value appears to be an outlier. (Middle) Plot of the logarithm (base 10) of the EC50. The distribution is now symmetrical. (Right) Plot of the raw data on a logarithmic axis. This kind of graph is a bit easier to read.

You can take 10 to a fractional power. Ten to the ½ power equals the square root of 10, which is 3.162. So the logarithm of 3.162 is 0.5.

Ten to the zero power equals 1, so the logarithm of 1.0 is 0.0.

You can take the logarithm of any positive number. The logarithms of values between zero and one are negative; the logarithms of values greater than one are positive. The logarithms of zero and negative numbers are undefined, because there is no power of 10 that results in a negative number or zero.

The *antilogarithm* (also called an antilog) is the inverse of the logarithm transform. Because the logarithm (base 10) of 1,000 equals 3, the antilogarithm of 3 is 1,000. To compute the antilogarithm of a base 10 logarithm, take 10 to that power.

All the examples above were about exponents of 10, so the logarithms defined above are base 10 logarithms, also called *common logarithms*. Logarithms can also be computed for other bases.

Why are logarithms useful?

Logarithms are useful because the logarithm of a product of two numbers equals the sum of the logarithm of the first value plus the logarithm of the second value. Logarithms convert multiplication into addition. This is the key insight that takes us back to lognormal distributions.

THE ORIGIN OF A LOGNORMAL DISTRIBUTION

Chapter 8 explained that a Gaussian distribution arises when variation is caused by many factors that are additive. Some factors push a value higher, some pull it lower, and the cumulative result is a symmetrical, bell-shaped distribution that approximates a Gaussian distribution.

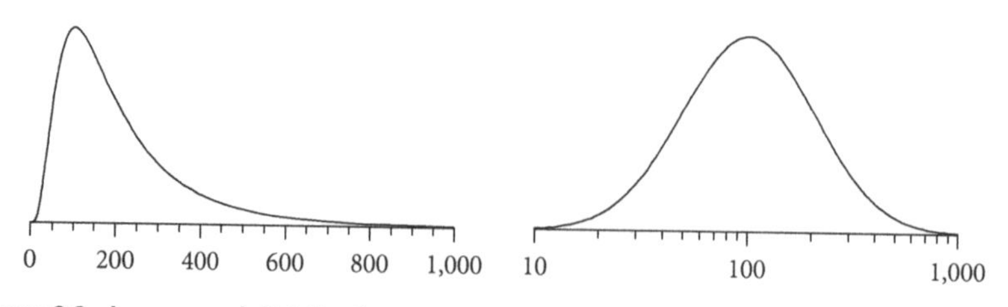

Figure 9.2. Lognormal distribution.
(Left) Lognormal distribution. (Right) Using a logarithmic X axis.

Some factors, however, act in a multiplicative rather than an additive manner. If a factor works multiplicatively, it is equally likely to double a value as to cut it in half. If that value starts at 100 and is multiplied by 2, it ends up at 200. If it is divided by 2, it ends up at 50. Consequently, that factor is equally likely to increase a value by 100 or to decrease it by 50. The effect is not symmetrical.

If many factors act in a multiplicative manner, the resulting distribution is asymmetrical, as shown in the graph on the left side of Figure 9.2. This distribution is called a *lognormal distribution*. Logarithms? How did logarithms get involved? Briefly, as explained above, it is because the logarithm of the product of two values equals the sum of the logarithm of the first value plus the logarithm of the second. So logarithms convert multiplicative scatter (lognormal distribution) to additive scatter (Gaussian distribution). Thus, the logarithms of the values sampled from a lognormal distribution follow a Gaussian distribution.

HOW TO ANALYZE LOGNORMAL DATA

If you transform each value sampled from a lognormal distribution to its logarithm, the distribution becomes Gaussian.

Let's revisit the example from the start the chapter. The middle graph in Figure 9.1 plots the logarithm of the EC50 values. Note that the distribution is now quite symmetrical.

The graph on the right in Figure 9.1 illustrates an alternative way to plot the data. The axis has a logarithmic scale. Note that every major tick on the axis represents a value 10 times higher than the previous value. The distribution of data points is identical to that in the middle graph, but the graph on the right is easier to comprehend because the Y values are in natural units of the data rather than logarithms.

GEOMETRIC MEAN

The mean of the data from Figure 9.1 is 1,333 nM. This is illustrated as a horizontal line in the graph on the left. The mean is larger than all but one of the values, so it is not a good measure of the central tendency of the data.

The graph in the middle of Figure 9.1 plots the logarithms of the values on a linear scale. The horizontal line is at the mean of the logarithms, or 2.71. About half of the values are higher, and about half are smaller.

The graph on the right in Figure 9.1 uses a logarithmic axis. The values are the same as those of the graph on the left, but the spacing of the values on the axis is logarithmic. The horizontal line is at the antilog of the mean of the logarithms. This graph uses logarithms base 10, so the antilog is computed by calculating $10^{2.71}$, which equals 513. The value 513 nM is called the *geometric mean*. For these data, this is a better measure of central tendency than the mean.

The geometric mean is expressed in the same units as the data.

To compute a geometric mean, first transform all the values to their logarithms (base 10), and then calculate the mean of those logarithms. Finally, transform that mean of the logarithms back to the original units of the data by taking 10 to that power. Some texts present an alternative way to compute the geometric mean: Multiply all the values together, and then take that product to the 1/n power, where n is the number of values.

LINGO

Don't mix up the geometric mean with the *arithmetic mean*, a fancy name for the ordinary mean (calculated by adding up all the numbers and divide by sample size).

COMMON MISTAKES

Mistake: Analyzing data from a lognormal distribution as if they were sampled from a Gaussian distribution

The results of any statistical analysis that assumes a Gaussian distribution are likely to be misleading when the data are actually lognormal. This is especially a problem with small samples.

Mistake: Identifying outliers without recognizing the data are lognormal

An easy mistake would be to view the graph on the left of Figure 9.1 and conclude that these values are sampled from a Gaussian distribution but contaminated by the presence of a single outlier. Running an outlier test would confirm that conclusion. But outlier tests assume that all the values (except for any outliers) are sampled from a Gaussian distribution. So when presented with data from a lognormal distribution, outlier tests are likely to incorrectly flag very high values as outliers when in fact those high values are expected in a lognormal distribution. Chapter 21 shows some examples.

Mistake: Trying to compute a geometric mean when some values are zero or negative

Recall that the logarithm is simply not defined when values are zero or negative. Therefore, the geometric mean can only be computed when every value is positive (greater than zero).

Q & A

Lognormal or log-normal?

Both forms are commonly used. This text uses lognormal.

Are values in a lognormal distribution always positive?

Yes. The logarithm of zero and negative numbers is simply not defined. Distributions that contain zero or negative values cannot be treated as lognormal distributions.

Can the geometric mean be computed if any values are zero or negative?

No.

When computing the geometric mean, should I use natural logarithms or logarithms base 10?

It doesn't matter, as long as you are consistent. More often, scientists use logarithms base 10, so the reverse transform is the power of 10. The alternative is to use natural logarithms, so the reverse transform is taking e to that power, where e is a math constant equal to 2.718. Whichever log base is used, the geometric mean will have exactly the same value.

Are lognormal distributions common?

Yes, they are very common in many fields of biology (Limpert, Stahel, & Abbt, 2001).

Are lognormal distributions always skewed to the right, as shown in the graph on the left of Figure 9.2?

Yes. The longer, stretched out tail always extends to the right, representing larger values.

What units are used to express the geometric mean?

The same units that are used with the values being analyzed. Thus, the mean and geometric mean are expressed in the same units.

Is the geometric mean larger or smaller than the regular mean?

The geometric mean is always smaller. (One trivial exception: If all the values are identical, then the mean and geometric mean will be equal.)

CHAPTER SUMMARY

- Lognormal distributions are very common in many fields of science.
- Lognormal distributions arise when multiple random factors are multiplied together to determine the value. In contrast, Gaussian distributions arise when multiple random factors are added together.
- Lognormal distributions have a long right tail (are skewed to the right).
- You may get misleading results if you choose analyses that assume sampling from a Gaussian distribution when your data are actually sampled from a lognormal distribution.
- In most cases, the best way to analyze lognormal data is to take the logarithm of each value and then analyze those logarithms.

CHAPTER 10

Confidence Interval for a Mean

INTERPRETING A CONFIDENCE INTERVAL FOR A MEAN

For our ongoing n = 130 body temperature example (see Figure 10.1), any statistics program will calculate that the 95% CI for the mean ranges from 36.75°C to 36.89°C. For the smaller n = 12 subset, the 95% CI for the mean ranges from 36.51°C to 37.02°C.

Note that there is no uncertainty about the sample mean. We are 100% sure that we have calculated the sample mean correctly. Any errors in recording the data or computing the mean will not be accounted for in computing the confidence interval of the mean. By definition, the confidence interval is always centered on the sample mean. The population mean is not known and cannot be known. However, given some assumptions discussed below, we can be 95% confident that the calculated interval contains it.

What, exactly, does it mean to be "95% confident"? When you have measured only one sample, you don't know the value of the population mean. The population mean either lies within the 95% CI or it doesn't. You don't know, and there is no way to find out. If you calculate a 95% CI from many independent samples, you expect that the population mean will be inside the confidence interval in 95% of the samples but outside the confidence interval in the other 5% of the samples. Therefore, using data from one sample, you can say that you are 95% confident that the 95% CI includes the population mean.

Confidence intervals can be written out as "the 95% CI ranges from 36.75 to 36.89" or [95% CI, 36.75–36.89]. Accompanying a legend that explains that they are 95% CI, they can also appear as "[36.75, 36.89]" or "36.82 ± 0.07".

WHAT VALUES DETERMINE THE CONFIDENCE INTERVAL FOR A MEAN?

The confidence interval for a mean is computed from four values:

- The sample mean. Our best estimate of the population mean is the sample mean. Accordingly, the confidence interval is centered on the sample mean.

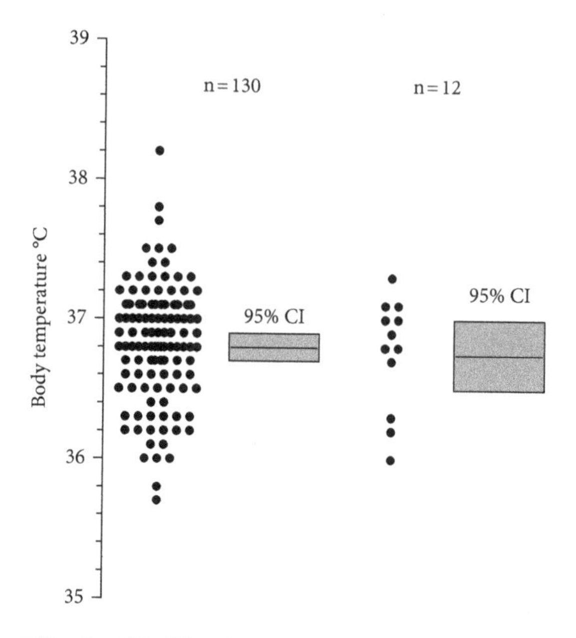

Figure 10.1. The 95% CI for the mean computed for the two sample data sets.

Note that the 95% CI does not contain 95% of the values, especially when the sample size is large.

- The standard deviation. If the data are widely scattered (large standard deviation), then the sample mean is likely to be farther from the population mean than if the data are tight (small standard deviation). The width of the confidence interval therefore is proportional to the sample standard deviation.
- The sample size. Our sample has 130 values. Therefore, the sample mean is likely to be quite close to the population mean, and the confidence interval will be very narrow. With tiny samples, the sample mean is likely to be farther from the population mean, so the confidence interval will be wider. The width of the confidence interval is inversely proportional to the square root of the sample size. If the sample were four times larger, the confidence interval would be half as wide (assuming the same standard deviation). Note that the confidence interval from the n = 12 sample is wider than the confidence interval for the n = 130 sample (see Figure 10.1).
- The degree of confidence. Although confidence intervals are typically calculated for 95% confidence, any value can be used. If you wish to have more confidence (e.g., 99% confidence), the confidence interval will be wider. If you are willing to accept less confidence (e.g., 90% confidence), the confidence interval will be narrower.

These four factors combine into this equation that defines the confidence interval of a mean:

$$\text{mean} - \frac{SD}{\sqrt{n}} \cdot t^* \quad \text{to} \quad \text{mean} + \frac{SD}{\sqrt{n}} \cdot t^*$$

The symbol t^* represents a constant whose value depends on the sample size and the confidence level you wish to have. You can compute that value using this Excel formula: =T.INV.2T(100-C,n-1), where C is 95 for 95% CIs, 90 for 90% CIs, etc. So if you want a 95% CI for a sample of 10 values, the formula would be =T.INV.2T(0.05,9) and t^* equals 2.262.

THE STANDARD ERROR OF THE MEAN

The equation above includes the ratio of the standard deviation divided by the square root of n. This value is called the *standard error of the mean*, abbreviated SEM in this book. That abbreviation is not entirely standard, partly because SEM has another meaning in advanced statistics: Structural Equal Modeling.

Often, the standard error of the mean is referred to simply as the *standard error*, with the word *mean* missing but implied (and abbreviated SE). Despite its name, the standard error of the mean has nothing to do with standards. How about errors? The mean you compute from a sample is unlikely to equal the true population mean (which you don't know), and that discrepancy is called the *sampling error*. The standard error of the mean is a way to quantify this sampling error. Standard errors can be computed for values other than the mean (e.g., the standard error of the best-fit value of a slope in linear regression; see Chapter 23).

You'll often see the standard error of the mean plotted as error bars on graphs, as discussed in the next chapter.

ASSUMPTIONS: CONFIDENCE INTERVAL FOR A MEAN

To interpret a confidence interval for a mean, you must accept the following assumptions, which are mostly the same as the assumptions for a confidence interval of a proportion presented in Chapter 4.

Assumption: Random (or representative) sample

The 95% CI is based on the assumption that your sample was randomly selected from the population. In many cases, this assumption is not strictly true. You can still interpret the confidence interval as long as you assume that your sample is representative of the population.

This assumption would be violated in the body temperature example if the participants chose to join the study because they knew (or suspected) that their own temperature was often higher or lower than that of most other people.

Assumption: Independent observations

The 95% CI is only valid when all subjects are sampled from the same population and each has been selected independently of the others. Selecting one member of the population should not change the chance of selecting any other member.

This assumption would be violated in the body temperature example if some individuals' temperatures were measured twice and both values were included in the sample. This assumption would also be violated if several of the subjects were siblings, because it is possible that genetic factors affect body temperature.

Assumption: Accurate data

The 95% CI is only valid when each value is measured correctly.

This assumption would be violated in the body temperature example if some people placed the thermometer in their mouths incorrectly or if the thermometer was misread.

Assumption: Assessing an event you really care about

The 95% CI allows you to use data from your sample to make an inference about the population. But the inference is only valid for the variable you measured even if you really care about a different variable.

In the body temperature example, what you really want to know is the core body temperature. Instead, temperature under the tongue was measured. The two may not be identical.

Chapter 27 will present two examples in which drugs changed a surrogate marker as desired—fewer arrhythmias in one example, higher "good cholesterol" in the other—but actually increased the number of deaths.

Assumption: The population is distributed in a Gaussian manner, at least approximately

The most common method for computing the confidence interval of a mean is based on the assumption that the data are sampled from a population that follows a Gaussian distribution. This assumption is important when the sample is small but doesn't matter much when samples are large.

What if the assumptions are violated?

In many situations, these assumptions are not strictly true. The patients in your study may be more homogeneous than the entire population of patients. Measurements made in one lab will have a smaller standard deviation than a set of measurements made in several labs at several times. More generally, the population you really care about may be more diverse than the population from which the data were sampled. Furthermore, the population may not be Gaussian. If any assumption is violated, the confidence interval will probably be too optimistic (too narrow). The true confidence interval (taking into account any violation of the assumptions) is likely to be wider than the calculated CI.

LINGO

It is easy to confuse confidence intervals and confidence limits. The two are very similar. A *confidence interval* is a range of values. The two values that define this range are called the *confidence limits*.

Distinguish the standard deviation from the standard error of the mean. The two are very different. The standard deviation quantifies scatter. The standard error of the mean quantifies precision – how close your sample mean is likely to be to the true population mean.

COMMON MISTAKES

Mistake: Thinking that a 95% CI covers the mean ± 2 SD

Nope! In a Gaussian distribution, you expect to find about 95% of the individual values within 2 SD of the mean. But the idea of a confidence interval is to define how precisely you know the population mean. For that, you need to take into account sample size.

Mistake: Thinking that a 95% CI covers the mean ± 2 SEM

With large samples, the confidence interval is approximated very well by the mean ± 2 SEM. But with small samples, the confidence interval is wider than that, and with tiny samples, it is much wider.

Mistake: Thinking that 95% of values lie within the 95% CI

The confidence interval quantifies how precisely you know the population mean. With large samples, you can estimate the population mean quite accurately, so the confidence interval is quite narrow and includes only a small fraction of the values (see Figure 10.1).

Mistake: Assuming that a result presented as 3.56 ± 1.23 is a confidence interval

As the next chapter explains, there are many possible meanings to data presented as a value plus or minus an error. Most likely, it is not a confidence interval.

Q & A

Why 95% confidence?

Confidence intervals can be computed for any degree of confidence. By convention, 95% CIs are presented most commonly, although 90% CIs and 99% CIs are sometimes published.

Is a 99% CI wider or narrower than a 95% CI?

To be more confident that your interval contains the population mean, the interval must be made wider. Thus, 99% CIs are wider than 95% CIs, and 90% CIs are narrower than 95% CIs (see Figure 10.2).

Does a confidence interval quantify variability?

No. The confidence interval of the mean tells you how precisely you have determined the population mean. It does not tell you about scatter among values.

Is the 95% CI of the mean always symmetrical around the mean?

It depends on how it is computed. But with the standard method, the confidence interval always extends equally above and below the mean.

Can I compute a confidence interval of the mean if I have only collected a single value?

No. Calculation of a confidence interval of the mean requires an assessment of variation, which requires more than one value.

Can I compute a confidence interval of the mean if I have only collected two values?

Yes (but you probably should collect more data).

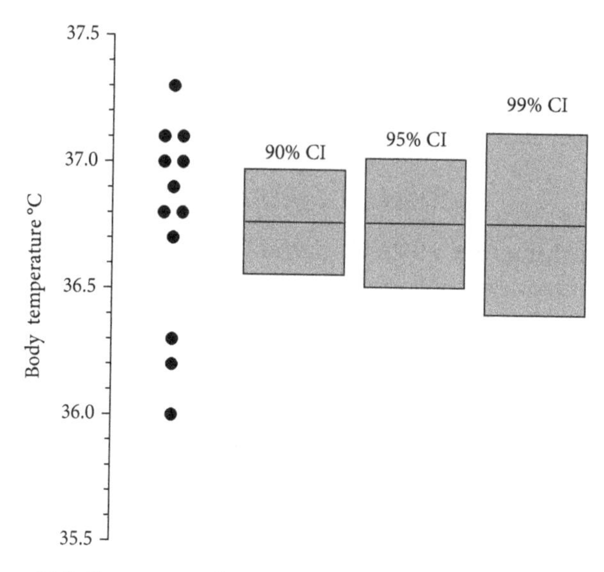

Figure 10.2. If you want to be more confident that a confidence interval contains the population mean, you must make the interval wider.
The 99% CI is wider than the 95% CI, which is wider than the 90% CI.

If I collect more data, should I expect 95% of the new values to lie within this confidence interval?

No! The confidence interval quantifies how precisely you know the population mean. With large samples, you can estimate the population mean quite accurately, so the confidence interval is quite narrow and includes only a small fraction of the values you sample in the future. This is shown in Figure 10.1.

CHAPTER SUMMARY

- A confidence interval of the mean shows you how precisely you have determined the population mean.
- If you compute 95% CIs from many samples, you expect that 95% of the samples will include the true population mean and that 5% will not. You'll never know whether a particular confidence interval includes the population mean.
- The confidence interval of a mean is computed from the sample mean, the sample standard deviation, and the sample size.
- The confidence interval does not display the scatter of the data. In most cases, the majority of the data values will lie outside the confidence interval.
- If you desire more confidence (99% rather than 95%), the confidence interval will be wider.
- Larger samples have narrower confidence intervals than smaller samples with the same standard deviation.
- Interpreting the confidence interval of a mean requires accepting a list of assumptions.

CHAPTER 11

Error Bars

THE APPEARANCE OF ERROR BARS

Figure 11.1 shows the maximal bladder relaxation (abbreviated % E_{max}) that can be achieved with large doses of norepinephrine in old rats (data from Frazier, Schneider, & Michel, 2006). You have already seen a different variable from this study in Chapter 9. The graph on the right shows the same data with several kinds of error bars representing the mean with the standard deviation, standard error of the mean, or 95% CI or the median with range or quartiles. The standard error of the mean was explained in the last chapter.

Figure 11.2 plots the mean with the standard deviation with or without horizontal caps and with bars extending in one or both directions. The choice among these styles is simply a matter of personal preference. When error bars with caps are placed on bars and only extend above the bar (as in Figure 11.2C), the resulting plot is sometimes called a *dynamite plunger plot*, sometimes shortened to *dynamite plot*.

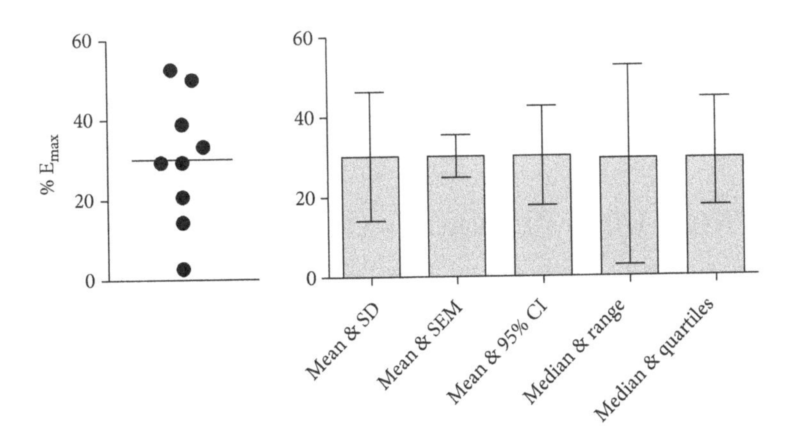

Figure 11.1. Five kinds of error bars.
(Left) The actual data. (Right) Five kinds of error bars, each representing a different way to portray variation or precision.

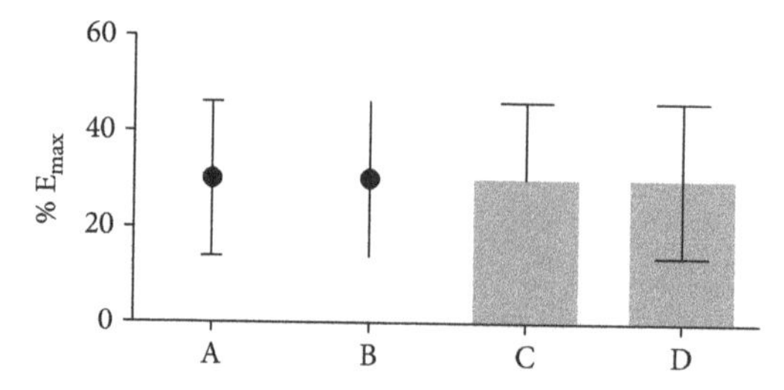

Figure 11.2. Four different styles of plotting the mean and standard deviation.

These methods all plot the same values. There is no real reason to prefer one style over another.

HOW TO INTERPRET ERROR BARS

Interpreting error bars is usually straightforward:

- **Standard deviation.** If the data are Gaussian, you expect about two-thirds of the values to lie within 1 SD of the mean and for about 95% to lie within 2 SD of the mean. Even if the data are not Gaussian, you can expect more than 75% of the values to lie within 2 SD of the mean.
- **Standard error of the mean.** The standard error of the mean was defined in the previous chapter. The best way to interpret these error bars is to compute the standard deviation and interpret that. Here is the needed equation, which is an exact calculation, not an approximation:

$$SD = SEM \cdot \sqrt{n}$$

- **95% CI.** Given some basic assumptions listed in the previous chapter, you know there is a 95% chance that the interval defined by the error bars includes the true population mean.
- **Quartiles.** You know that half of the observed values are between the 25th and the 75th percentile.
- **Range.** You know that all of the observed values lie in that range. Most future values will also probably lie in that range, but some may not.

WHICH KIND OF ERROR BAR SHOULD YOU PLOT?

An *error bar* displays variation or uncertainty. Your choice of error bar depends on your goals.

Goal: To show the variation among the values

If each value represents a different individual, you probably want to show the variation among the values. Even if each value represents a different lab experiment, it often makes sense to show the variation.

With fewer than 100 or so values, you can create a scatter plot that shows every value. What better way to show the variation among values than to show every value? But if your data set has more than 100 or so values, a scatter plot becomes messy. Alternatives include a box-and-whiskers plot and a frequency distribution (histogram) (see Chapter 6).

What about plotting the mean and the standard deviation? The standard deviation does quantify scatter, so this is indeed one way to graph variability. But the standard deviation is only one value, so plotting it is a pretty limited way to show variation. A graph showing the mean and SD error bars is less informative than any of the other alternatives. The advantages of plotting a mean with SD error bars (as opposed to a column scatter graph, a box-and-whiskers plot, or a frequency distribution) are that the graph is simpler and the approach is conventional in many fields.

Of course, if you do show SD error bars, be sure to say so in the figure legend. Otherwise, they might be confused with SEM error bars.

If you are creating a table rather than a graph, there are several ways to present the data, depending on how much detail you want to show. One choice is to show the mean as well as the smallest and largest values. A more compact choice is to tabulate the mean plus or minus the standard deviation.

Goal: To show how precisely you have determined the population mean

Perhaps your goal is to show how precisely the data define the population mean—in other words, to show the likely sampling error. If this is your goal, I think the best approach is to plot the 95% CI of the mean.

What about the standard error of the mean? Graphing the mean with SEM error bars is a method commonly used to show how precisely you have determined the population mean. The only advantages of SEM error bars versus CI error bars is that SEM error bars are shorter and, in some fields, are more conventional. However, it is harder to interpret SEM error bars than it is to interpret CI error bars.

Whatever error bars you show, be sure to state your choice.

Goal: To create persuasive propaganda

Simon (2005) "advocated" the following approach in his excellent blog. Of course, he meant it as a joke!

- When you are trying to emphasize small and unimportant differences in your data, show your error bars as the standard error of the mean and hope that your readers think they are the standard deviation. (The readers will believe that the standard deviation is much smaller than it really is and so will incorrectly conclude that the difference among means is large compared to the variation among replicates.)
- When you are trying to cover up large differences, show the error bars as the standard deviation and hope that your readers think they are the standard error of the mean. (The readers will believe that the standard deviation is much larger than it really is and so will incorrectly conclude that the difference among means is small compared to the variation among replicates.)

HOW ARE STANDARD DEVIATION AND STANDARD ERROR OF THE MEAN RELATED TO SAMPLE SIZE?

If you increase the sample size, is the standard deviation expected to get larger, get smaller, or stay about the same?

For the body temperature example, the standard deviation is 0.41°C for the n = 130 data set and 0.40°C for the n = 12 sample. Many find it surprising that the standard deviation is so similar in samples of such different sizes. But in fact, this is to be expected. The standard deviation quantifies the scatter of the data. Whether increasing the size of the sample increases or decreases the scatter in that sample depends on the random selection of values. The standard deviation therefore is equally likely to get larger or to get smaller as the sample size increases. As you collect larger samples, you'll be able to quantify the variability (the standard deviation) more precisely, but collecting more data won't change the variability among the values.

If you increase the sample size, is the standard error of the mean expected to get larger, get smaller, or stay about the same?

For the body temperature example, the standard error of the mean is 0.1157°C for the n = 12 sample and only 0.0359°C for the n = 130 data set. This makes sense. The standard error of the mean quantifies how precisely the population mean has been determined. A larger sample usually has a smaller standard error of the mean than a smaller sample does, because the mean of a large sample is likely to be closer to the true population mean than the mean of a small sample is.

Of course, in any particular experiment, you can't be sure that increasing the sample size will decrease the standard error of the mean. It is a general rule, but it is also a matter of chance. So it is possible, but unlikely, that the standard error of the mean will increase if you collect a larger sample size.

LINGO

The term *error* can be confusing as error bars don't display error in the usual sense of that word (to indicate mistakes were made). Instead error bars display either variation among values or the precision of the mean (or perhaps median).

COMMON MISTAKES

Mistake: Plotting mean and error bar instead of plotting a frequency distribution

Figure 11.3 shows the number of citations (references in other papers within five years of publication) received by 500 papers randomly selected from the journal *Nature* (Colquhoun, 2003). With a glance at the graph, you can see that many papers are only cited a few times (or not at all), while only a few are heavily cited.

The mean number of citations is 115, with a standard deviation of 157 and a standard error of the mean of 7. Viewing a graph showing that the mean is 115

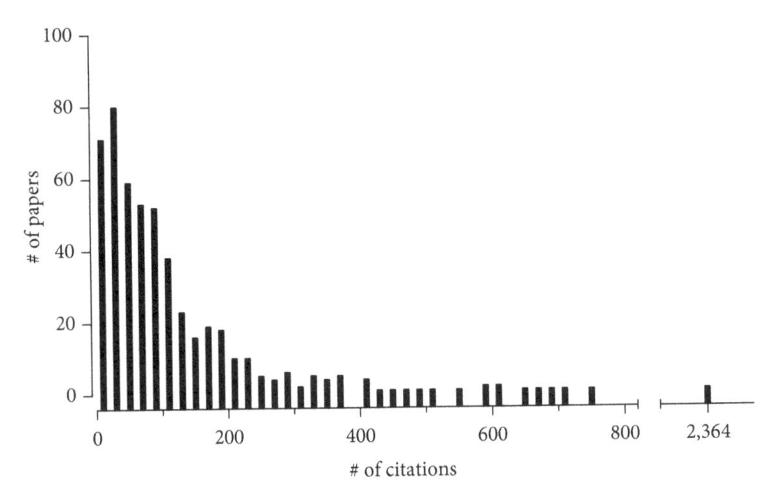

Figure 11.3. A very asymmetrical distribution.
This figure, remade from Colquhoun (2003), shows the number of citations (X axis) received within five years of publication by 500 papers randomly chosen from the journal *Nature*. The bin width is 20 citations. The first bar shows that 74 papers received between 0 and 20 citations, the second bar that 80 papers received between 21 and 40 citations, and so on. With such a skewed distribution, you really need to see the frequency distribution to understand the data. Knowing the mean and the standard deviation (or the standard error of the mean) is not enough.

and the standard deviation is 157 (or the standard error of the mean is 7) would not be very helpful in understanding the data.

The median number of citations is approximately 65, and the interquartile range runs from about 20 to 130. Reporting the median with that range would be more informative than reporting the mean and the standard deviation or standard error of the mean. However, it seems to me that the best way to summarize these data is to show a frequency distribution. No numerical summary really does a good job.

Mistake: Assuming that all distributions are Gaussian

What would you think about the distribution of a data set if you were told only that the mean is 101 and the standard deviation is 43—or if you saw the bar graph of Data Set A in Figure 11.4? You'd probably imagine that the data look something like Data Set B in the figure, and that would often be correct. But Data Sets C and D of Figure 11.4 also share that same mean and standard deviation (and standard error of the mean). The distributions of values in Data Sets C and D are quite different from the distribution of values in Data Set B, but all three data sets have the same mean, standard deviation, and standard error of the mean. It can be useful to summarize data as the mean and standard deviation (or standard error of the mean), but it can also be misleading. Look at the raw data (or a box-and-whiskers plot) when possible.

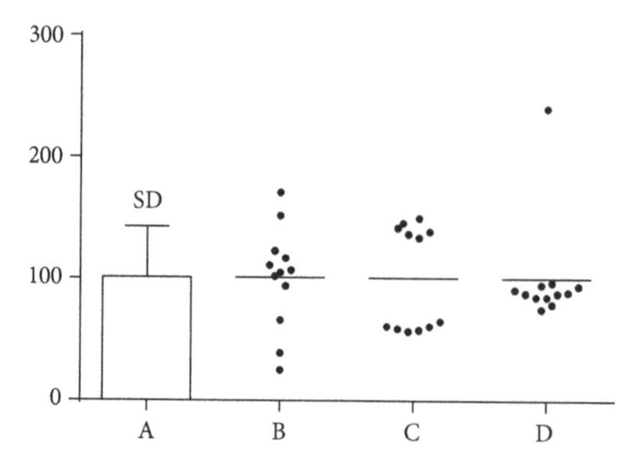

Figure 11.4. The mean and standard deviation can be misleading. All four data sets in the graph have approximately the same mean (101) and standard deviation (43). If you were only told those two values or only saw the bar graph of Data Set A, you'd probably imagine that the data look like Data Set B. But Data Sets C and D, with very different distributions, also have the same mean, standard deviation, and standard error of the mean as Data Set B.

Mistake: Plotting means and error bars without defining how the error bars were computed

If you make a graph showing means and error bars (or a table with means plus or minus error values), it is essential that you state clearly what the error bars are—standard deviation, standard error of the mean, confidence interval, range, or something else. Without a clear legend or a statement in the methods section, the graph or table is ambiguous.

Mistake: Plotting SEM error bars simply because they are the shortest

SEM error bars are popular because they are always smaller than the standard deviation. The main problem with SEM error bars is that some people mistake them for SD error bars and think they show the variation among the data.

Q & A

What does the standard deviation quantify?

The standard deviation quantifies scatter—in other words, how much the values vary from one another.

What does the standard error of the mean quantify?

The standard error of the mean quantifies how precisely you know the true mean of the population. The value of the standard error of the mean depends upon both the standard deviation and the sample size.

Are the standard deviation and the standard error of the mean expressed in the same units?

Yes. Both are expressed in the same units as the data.

Which is smaller?

The standard error of the mean is always smaller than the standard deviation.

Graphs often show an error bar extending 1 SEM in each direction from the mean. What does this denote?

The range defined by the mean ± 1 SEM has no simple interpretation. With large samples, that range is a 68% CI of the mean. But the degree of confidence depends on sample size. When n = 3, that range is only a 58% CI. You can think of the mean ± 1 SEM as covering about a 60% CI.

CHAPTER SUMMARY

- The standard deviation quantifies variation.
- The standard error of the mean quantifies how precisely you know the population mean. It does not quantify variability. The standard error of the mean is computed from the standard deviation and the sample size.
- The standard error of the mean is always smaller than the standard deviation. Both are measured in the same units as the data.
- Graphs are often plotted as means and error bars, which are usually either the standard deviation or the standard error of the mean.
- Consider plotting each value, or a frequency distribution, before deciding to graph a mean and an error bar.
- If you see a graph or table showing the standard error of the mean, multiply it by the square root of sample size to calculate the standard deviation.
- Use the standard deviation when your goal is to show variability, and use the standard error of the mean (or, better, the confidence interval) when your goal is to show how precisely you have determined the population mean.
- Graphs with error bars (or tables with error values) should always contain labels or a legend explaining whether the error bars are the standard deviation, the standard error of the mean, or something else.

CHAPTER 12

Comparing Groups with Confidence Intervals

USING CONFIDENCE INTERVALS TO COMPARE GROUPS

Previous chapters explained the concept of using a confidence interval to quantify how precisely you have determined a proportion or mean from a sample of data. When comparing two treatments, you can quantify the precision of the effect of the treatment, often a difference or ratio.

What then? Is the treatment effect large or small? Is the confidence interval too wide to be useful, or is it narrow enough to lead to reasonably precise conclusions? The examples below demonstrate that these questions cannot be answered by statistics. Scientists answer those questions, after computing the confidence intervals, by using common sense combined with a deep understanding of the scientific or clinical context and the limitations of the study design.

EXAMPLES OF CONFIDENCE INTERVALS USED TO COMPARE GROUPS

Example 1: Does apixaban prevent thromboembolism?

Agnelli and colleagues (2013) tested whether apixaban (a new anticoagulant or "blood thinner") would be beneficial in extended treatment of patients with a venous thromboembolism (a clot in the leg vein), which can be dangerous if it breaks loose and goes to the lung. After all the patients completed the standard anticoagulation therapy, the investigators randomly divided them into two groups that were assigned to receive either placebo or apixaban. The investigators followed these patients for a year and recorded whether they experienced another thromboembolism.

The thromboembolism recurred in 8.8% of the patients receiving the placebo (73/829) but in only 1.7% of the people receiving apixaban (14/840). So patients who received apixaban were a lot less likely to have a recurrent thromboembolism. The drug worked!

As explained in Chapter 4, we could compute the confidence interval for each of those proportions (8.8% and 1.7%), but the results are better summarized by computing the ratio of the two proportions and its confidence interval. This ratio is termed the *risk ratio*, or the *relative risk*. The ratio is 1.7/8.8, which is 0.19. Patients receiving the drug had 19% the risk of a recurrent thromboembolism compared to patients receiving the placebo.

That ratio is an exact calculation for the patients in the study. A 95% CI generalizes the results to the larger population of patients with thromboembolism similar to those in the study. This book won't explain how the calculation is done, but many programs will report the result. The 95% CI of this ratio extends from 0.11 to 0.33. If we assume the patients in the study are representative of the larger population of adults with thromboembolism, then we are 95% sure that treatment with apixaban will reduce the incidence of disease progression to somewhere between 11% and 33% of the risk in untreated patients. In other words, patients in the study taking apixaban had about one-fifth the risk of another clot than those taking placebo did, and if we studied a much larger group of patients (and accepted some assumptions), we could be 95% confident that the risk would be somewhere between about one-ninth and one-third the risk of those taking placebo.

Is reducing the risk of thromboembolism to one-fifth its control value a large, important change? That is a clinical question, not a statistical one, but I think anyone would agree that is a substantial effect. The 95% CI goes as high as 0.33. If that were the true effect, a reduction in risk down to only one-third of its control value, it would still be considered substantial. When interpreting any results, you also have to ask whether the experimental methods were sound (I don't see any problem with this study) and whether the results are plausible (yes, it makes sense that extending treatment with a blood thinner might prevent thromboembolism). Therefore, we can conclude with 95% confidence that apixaban substantially reduces the recurrence of thromboembolism in this patient population. (Note that I have avoided the word "significant", which will be explained in Chapter 14.)

Example 2: Is the maximal bladder relaxation different in old and young rats?

Frazier, Schneider, and Michel (2006) compared old and young rats to see how well the neurotransmitter norepinephrine relaxes bladder muscles. Figure 12.1 shows the maximal relaxation that can be achieved by large concentrations of norepinephrine in old and young rats. These values are the percentage of relaxation, where the scale from 0% to 100% is defined using suitable experimental controls.

The values for the two groups of rats overlap considerably, but the two means are distinct. The mean maximum response in old rats is 23.5% lower than the mean maximum response in young rats. That value is exactly correct for our sample of data, but we know the true difference in the populations is unlikely to equal 23.5%. To make an inference about the population of all similar rats, look at the 95% CI of the difference between means. This is done as part of the calculations of a statistical test called the *unpaired t test* (not explained in detail in this

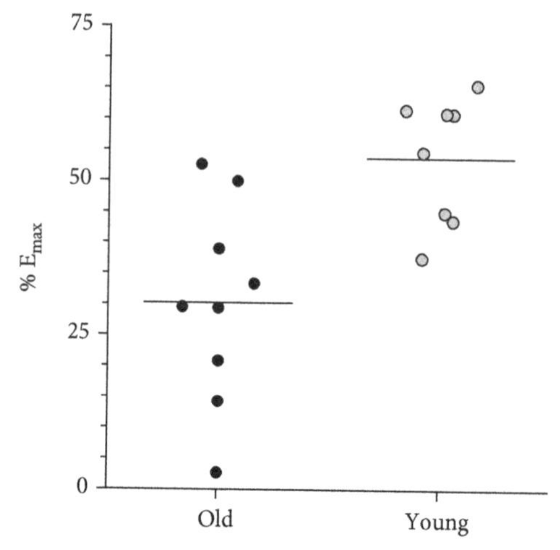

Figure 12.1. Maximal relaxation of muscle strips of old and young rat bladders stimulated with high concentrations of norepinephrine. More relaxation is shown as larger numbers. Each symbol represents a measurement from one rat. The horizontal lines represent the means. The symbols are nudged right or left so that they don't overlap, but otherwise, the horizontal position has no meaning.

book). This confidence interval (for the mean value measured in the young rats minus the mean value observed in the old rats) ranges from 9.3 to 37.8, and it is centered on the difference we observed in our sample. Its width depends on the sizes (number of rats) and the variability (standard deviation) of the two samples and on the degree of confidence you want (95% is standard).

Because the 95% CI does not include zero, we can be at least 95% sure that the mean response in old rats is less than the mean response in young ones. Beyond that, the interpretation needs to be made in a scientific context. Is a difference of 23.5% physiologically trivial or large? How about 9.3%, the lower limit of the confidence interval? These are not statistical questions. They are scientific ones. The investigators concluded this effect is large enough (and is defined with sufficient precision) to perhaps explain some physiological changes with aging.

Example 3: Does prednisolone extend survival in people with chronic active hepatitis?

Figure 12.2 shows survival of patients with chronic active hepatitis who were treated with either prednisolone (a steroid) or placebo (Kirk et al., 1980; raw data from Altman & Bland, 1998). Some patients were still alive at the time the data were collected, so the curves end up with survival well above 0%. This book does not explain survival curves, but you can read an overview in Chapters 5 and 29 of the larger version of this text (Motulsky, 2014).

One way to summarize the data is to compute the median survival of each group, defined as the time it takes until half the subjects have died. The horizontal line in the graph on the right of Figure 12.2 is at 50% survival. The

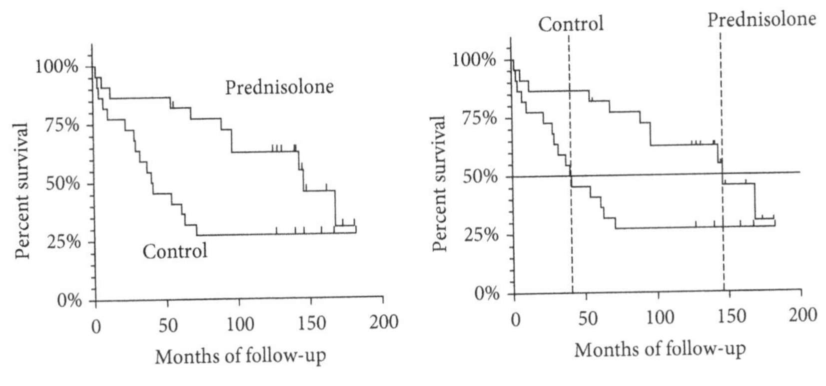

Figure 12.2. Survival curves.
(Left) Percent survival as a function of time, comparing two groups of 22 patients with chronic active hepatitis. (Right) Median survival times of both groups.

time at which each survival curve crosses that line is the median survival. The median survival of the control patients is 40.5 months, whereas the median survival of the patients treated with prednisolone is 146 months. The ratio is 3.6. The median survival of patients treated with prednisolone is 3.6 times the median survival of the control patients.

This ratio (3.6) exactly describes the survival of the 44 patients in the study. To generalize to the larger group of patients with chronic active hepatitis, look at the 95% CI for that ratio, reported to range from 3.1 to 4.1. In other words (and with some approximation), we are 95% sure that treatment with prednisolone has an effect somewhere between tripling and quadrupling the median survival time.

Is increasing median survival by a factor of 3.6 a clinically relevant, large change? That is a clinical question, not a statistical one, but clearly, the answer is yes. The lower limit of the 95% CI is 3.1. Is it clinically important to know that a drug triples median survival time? Again, I think most would agree that it is. Therefore, the results of this study clearly show that the drug substantially increases survival and that the extent of this increase is determined with reasonable precision.

Example 4: Do selective serotonin reuptake inhibitors increase the risk of autism?

Hviid and colleagues (2013) asked whether there is an association between use of selective serotonin reuptake inhibitors (SSRIs, a treatment for depression) by pregnant women and autism in their children. They followed 626,875 babies from birth and identified 3,892 kids diagnosed with autism and 6,068 mothers who had used SSRIs during pregnancy. They used a fairly sophisticated method to analyze the data, but it is easy to understand the results without understanding the details of the methodology. The risk of autism was 1.2 times higher among children of women who took SSRIs compared to the risk in children of those who didn't.

To generalize to the larger population, the investigators computed a 95% CI for that risk ratio, which ranged from 0.9 to 1.6. A ratio of 1.0 would mean that the risk of having a child with autism is the same in women who took SSRIs as in those who didn't. The confidence interval extends from a risk of less than 1.0 (which would mean SSRIs are associated with a lower risk of autism) to a value of greater than 1.0 (which would mean SSRIs are associated with a higher risk of autism).

Clearly, these results do not demonstrate an association between use of SSRIs during pregnancy and autism. But do they convincingly show that there is no association? Not really. The 95% CI goes up to 1.6, which represents a 60% increase in the risk of autism. That is a lot of risk. The investigators therefore do not give a firm negative conclusion but instead conclude (p. 2406), "On the basis of the upper boundary of the confidence interval, our study could not rule out a relative risk up to 1.6, and therefore the association warrants further study."

Example 5: Does tight hyperglycemic control benefit children in intensive care?

When hospitalized patients have hyperglycemia (high blood sugar), it is hard to know what to do. Administering insulin can maintain blood glucose at normal levels, but this creates a danger of hypoglycemia (too little glucose in the blood), which has negative consequences. Macrae and colleagues (2014) did a controlled study asking whether tight hyperglycemic control (more insulin) benefited children in intensive care. They randomized 1,369 patients to receive either tight or conventional glucose control. The primary outcome they tabulated was the number of days alive and not on a mechanical ventilator. On average, the patients who received the tighter control of hyperglycemia remained alive (and off a ventilator) 0.36 days longer than patients given standard glucose control.

To generalize their findings to the larger population of children in intensive care, they presented the 95% CI, which ranged from −0.42 to 1.14 days. The 95% interval includes zero, so we can say with 95% confidence there is no evidence that the tighter glucose control was beneficial. But is this confidence interval narrow enough that these findings can be considered a solid negative result? The authors stated that they would have considered a difference of two days to be clinically relevant. The upper limit of the confidence interval was only a bit more than one day. So we can conclude that these results show the difference between the two treatments, at best, is not relevant clinically. These are solid negative results. (The investigators analyzed the data in several ways and looked at subgroups of patients. The final conclusions are a bit more nuanced than those presented above.)

Example 6: Does eating processed meat increase the risk of colon cancer?

Larsson and Wolk (2006) pooled results from 14 published studies (including more than a million people) to determine whether people who eat processed meat have a higher risk of colon cancer. A formal study like this that quantitatively combines results from previously published studies is called a *meta-analysis*. These investigators compared the incidence of colon cancer in the people who reported the most processed meat consumption with the incidence in those who reported the least. The relative risk ratio of colon cancer was 1.20, with a higher risk in those who ate more processed meat. In other words, the risk of cancer was 20% higher in people who reported eating processed meat. This does not mean that 20% of those people had cancer, only that the incidence of cancer in that group was 1.20 times the incidence in the comparison group.

To extrapolate to the general population, they reported the 95% CI of the relative risk, which ranges from 1.11 to 1.31. Since the confidence interval does not include 1.0 (which would indicate no increased risk), we can say with 95% confidence that people who report higher consumption of processed meat are more likely to get colon cancer.

Evaluating these data requires looking at the magnitude of the effect. An elevated risk of 20% is certainly somewhat worrying, and this is how the authors discussed the finding in the discussion section of their paper. But a 20% increase (or even a 31% increase, the upper confidence limit) in cancer incidence is still fairly small. Is this difference really due to consumption of processed meats? None of the studies pooled here was an actual experiment. People were not randomized to eat more or less processed meat. People ate whatever they wanted and filled out questionnaires about their eating habits. The problem is that people who eat a lot of processed meat are probably different in many ways from people who eat little processed meat. For example, people who eat more processed meat may smoke and drink more and be heavier. The epidemiologists who performed the studies tried to correct for these other differences, called *confounding variables*, but these corrections can't be perfect. Experienced epidemiologists say that you really can't believe associations from this kind of study unless the relative risk is greater than 2.0 or 3.0 (UCSF Legacy Tobacco Documents Library, 2008). So it certainly is reasonable to conclude that these results are inconclusive even though the confidence interval does not include 1.0 (Colquhoun, 2013). Note that interpreting these results depends on more than reviewing the statistical calculations. It also requires understanding of the scientific context and the limitations of the study design.

ASSUMPTIONS OF CONFIDENCE INTERVALS

Chapter 4 discussed the assumptions required to interpret a confidence interval of a proportion, and the same assumptions apply when comparing two groups:

- Random or representative sample.
- Independent observations.
- Accurate data.

Additionally, each statistical calculation may introduce more assumptions. For instance, the second example in this chapter assumes that the values are sampled from a population in which the values follow a bell-shaped Gaussian distribution, at least approximately. The examples with survival curves makes additional assumptions, which are not detailed in this book.

COMMON MISTAKES

Mistake: Interpreting the confidence interval as a description of your data

The confidence interval is used to make a general conclusion about the population from which the data were sampled. The confidence interval doesn't describe the data you collected.

Mistake: Interpreting the confidence interval when the assumption of a random or representative sample is badly violated

Since the confidence interval makes a general conclusion from limited data, it depends on assumptions. The main assumption is that your data are randomly sampled from (or at least representative of) a larger population for which you are making a generalized conclusion.

Mistake: Focusing only on whether the confidence interval includes the value that means no effect

It is worth noting whether the confidence interval includes the value that indicates no effect (a difference of 0.0, or a ratio of 1.0). But it is also important to look at how wide the confidence interval is and interpret both ends of the interval in terms of scientific or clinical impact.

Mistake: Instead of looking at the confidence interval for the difference (or ratio), looking separately at the confidence interval for the two groups and asking if they overlap

You learn less than you'd expect to by asking whether two error bars overlap. Instead, base your conclusions on the confidence interval for the difference or ratio.

Q & A

Why is there no mention of P values or statistical significance in any of these examples?

These examples show that confidence intervals are a very useful way to compare groups. The next chapter (13) will explain P values, and the one after (14) that will explain statistical significance.

CHAPTER SUMMARY

- Confidence intervals can be used in any situation where a comparison can be expressed as a single value, usually a difference or ratio, and you want to make an interpretation for the entire population, not just the sample of data actually analyzed.

- Even if you don't understand the details of a statistical method, you can understand the meaning of the reported 95% CI.
- Making conclusions from data requires understanding the scientific context and the questions the experiment or study was designed to answer.
- All confidence intervals—indeed, all statistical inferences— are based on assumptions.

CHAPTER 13

Comparing Groups with P Values

INTRODUCING P VALUES VIA COIN FLIPPING

You flip a coin 20 times and observe 16 heads and 4 tails. Because the probability of heads (using a fair coin) is 50%, you'd expect about 10 heads in 20 flips. How unlikely is it to find 16 heads in 20 flips? Should you suspect that the coin is unfair?

Defining the null hypothesis

To evaluate our suspicion that the coin might be unfair, we need to define a fair coin. A coin flip is fair when each coin flip has a 50% chance of landing on heads and a 50% chance of landing on tails and when each result is accurately recorded. For the purposes of computing a P value, this is called the *null hypothesis*. In this example, the null hypothesis is that the coin tosses are fair. In other situations, as you'll see below, the null hypothesis is that the ratio really is 1.0 (no effect) or the difference really is 0.0 (no change).

Calculating the P value

Assuming the null hypothesis is true, what probability do we want to compute? We could compute the chance of observing 16 heads in 20 flips, but that isn't sufficient. We would have been even more suspicious that the coin flipping wasn't fair had the coin landed on heads 17 times, or 18 times, and so on. Similarly, we would have been suspicious if the coin had landed on tails 16 or more times (and on heads 4 or fewer times). Putting all this together, we want to answer the following question:

> If the coin tosses are random and the answers are recorded correctly, what is the chance that when you flip the coin 20 times, you'll observe either 16 or more heads or 4 or fewer heads?

In other words, we need to sum the probability of seeing 0, 1, 2, 3, 4, 16, 17, 18, 19, or 20 heads in 20 coin flips. This book won't explain how to do this calculation, but the answer is 0.0118, or 1.18%. Put another way, you would only see results this far from an even split of heads to tails in 1.18% of 20-coin runs.

This result is called a *P value*. It is almost always reported as a decimal fraction (0.0118) rather than a percentage.

Interpreting the P value

Should you conclude that the coin is unfair? There are two possible explanations for your observations. One is that they are a coincidence. The other is that the coin flipping wasn't fair. Which is it? All statistics can do is tell you how rare it is that such a coincidence would happen (1.18% of the time). What you conclude depends on the context. Consider these three scenarios (already mentioned in Chapter 4):

- You've examined the coin and are 100% certain it is an ordinary coin. You also did the coin flips and recorded the results yourself, so you are sure there is no trickery. In this case, you can be virtually 100% certain that this streak of many heads is just a chance event. The calculation shown above won't change that conclusion.
- If the coin flips were part of a magic show, you can be pretty sure that trickery invalidates the assumptions that the coin is fair, the tosses are random, and the results are accurately recorded. You'll conclude that the coin flipping was not fair.
- If everyone in a class of 200 students had flipped the coin 20 times and the run of 16 heads was the most heads obtained by any student, you wouldn't conclude that the coin was unfair. It is not surprising to see a coincidence that happens 1.18% of the time when you run the test (coin flipping) 200 times. You'd expect that to happen at least once.

A RULE THAT LINKS CONFIDENCE INTERVALS AND P VALUES

In almost all cases, a statistical analysis can report both a confidence interval and a P value. Here are some general rules that link the two for most analyses:

- If a 95% CI does not contain the value of the null hypothesis, then the P value must be less than 0.05. For example, if the confidence interval for the difference between two means does not include zero, or if the confidence interval for the ratio of two proportions does not include 1.0, then the P value must be less than 0.05.
- If a 95% CI does include the value of the null hypothesis, then the P value must be greater than 0.05.
- Similarly, if the 99% CI does not contain the null hypothesis value, then the P value must be less than 0.01. If a 90% CI does not contain that value, then the P value must be less than 0.10.

REVISITING THE EXAMPLES FROM CHAPTER 12

The previous chapter presented examples of comparing groups using 95% CIs. Now, let's interpret P values for those same examples.

Example 1: Does apixaban prevent thromboembolism?

The previous chapter presented the example of apixaban used to prevent thromboembolism (Agnelli et al., 2013). Those investigators reported a risk ratio of 0.19 along with its 95% CI of 0.11 to 0.33.

Could the results be due to a coincidence of random sampling? To assess this question, we first must tentatively assume that the risk of thromboembolism is the same in patients who receive an anticoagulant as in those who receive placebo and that the discrepancy in the incidence of thromboembolism observed in this study was the result of chance. This is the null hypothesis. Then, we can ask the following question:

> If the risk of thromboembolism overall is identical in the two groups, what is the chance that random sampling (in a study as large as ours) would lead to a ratio of incidence rates as far or farther from 1.0 as the ratio computed in this study? (Why 1.0? Because a ratio of 1.0 implies no treatment effect.)

The authors did the calculation using a test called the *Fisher exact test* and published the answer as $P < 0.0001$. When P values are tiny, it is traditional to simply state that they are less than some small value—here, 0.0001. Since this P value is low, and since the hypothesis that an anticoagulant would prevent thromboembolism is quite sensible, the investigators concluded that the drug worked to prevent recurrent thromboembolism.

Example 2: Is the maximal bladder relaxation different in old and young rats?

Frazier, Schneider, and Michel (2006) compared maximal relaxation of bladder muscle when exposed to norepinephrine between old and young rats (Figure 12.1). The null hypothesis is that both sets of data are randomly sampled from populations with identical means. The P value answers the following question:

> If the null hypothesis is true, what is the chance of randomly observing a difference as large as or larger than the one observed in this experiment of this size?

The P value depends on the difference between the means, on the standard deviation of each group, and on the sample sizes. A test called the *unpaired t test* (also called the *two-sample t test*) reports that the P value equals 0.003. If old and young rats have the same maximal relaxation of bladder muscle by norepinephrine overall, there is only a 0.03% chance of observing such a large (or a larger) difference in an experiment of this size.

Example 3: Does prednisolone extend survival in people with chronic active hepatitis?

The third example compared the survival curves of patients with chronic active hepatitis treated with prednisolone versus placebo (Kirk et al., 1980; raw data from Altman & Bland, 1998).

When comparing two survival curves, the null hypothesis is that the two survival curves are identical in the overall population of patients and the difference observed in the experimental samples was simply the result of chance. The P value answers the following question:

> If the null hypothesis is true, what is the probability of randomly selecting subjects whose survival curves are as different as or more different than what was actually observed?

The P value, computed by a test called the *log-rank test,* is 0.031. If the treatment really is ineffective, it is possible that the patients who were randomly selected to receive one treatment just happened to live longer than the patients who received the other treatment, with survival curves as far apart as observed in the study. The P value tells us that the chance of this happening is only 3.1%.

Example 4: Do selective serotonin reuptake inhibitors increase the risk of autism?

Hviid and colleagues (2013) compared the incidence of autism in children born to mothers who took selective serotonin reuptake inhibitors compared to those who didn't. The P value answers the following question:

> If the risk of autism is identical in children of mothers who took selective serotonin reuptake inhibitors and in those of mothers who didn't, what is the chance that random sampling will result in a relative risk as far or farther from 1.0 as observed in this study?

The authors reported a 95% CI for the relative risk ratio but did not report a P value. Since the 95% CI of the risk ratio includes 1.0 (the null hypothesis signifying no change in risk), we know the P value must be greater than 0.05, but we don't know by how much.

Example 5: Does tight hyperglycemic control benefit children in intensive care?

Macrae and colleagues (2014) compared the incidence of death or need for a ventilator in hospitalized children treated with two different protocols to manage hyperglycemia. The P value answers the following question:

> If the risk of death or need for a ventilator is identical among children in intensive care treated with the two alternative protocols, what is the chance that random sampling in a study of this size will result in a relative risk as far or farther from 1.0 as observed in this study?

The author reported a 95% CI, but not a P value, of the primary finding. Since the 95% CI included zero (no difference, the null hypothesis), we know that the P value has to be greater than 0.05.

Example 6: Does eating processed meat increase the risk of colon cancer?

Larsson and Wolk (2006) pooled results from 14 published studies (including more than a million people) to determine whether people who eat processed meat have a higher risk of colon cancer. The P value answers the following question:

> If the risk of colon cancer is identical in people who report high and low consumption of processed meat, what is the chance that random sampling in a study of this size will result in a relative risk as far or farther from 1.0 as observed in this study?

As mentioned, the investigators did not report P values, but they did report the 95% CI of the relative risk. Since the 95% CI of the relative risk does not include 1.0, we can be sure that the P value must be less than 0.05.

FOUR THINGS YOU NEED TO KNOW ABOUT P VALUES

Sample size matters when interpreting P values

A P value quantifies the probability of observing a difference or association as large as or larger than actually observed if the null hypothesis is true. Table 13.1 demonstrates that the relationship between the P value and the size of the observed effect depends on sample size. When reviewing scientific results, it is essential to look at the size of the effects and not just at the P value.

The logic of P values can seem weird, even backward

The logic of P values goes in a direction you may not expect. In conducting a study, you observe a sample of data and ask about the populations from which the data are sampled. The definition of the P value starts with an assumption about the population (the null hypothesis) and asks about possible data samples. Thinking about P values may seem quite counterintuitive to you—unless you are a lawyer or Talmudic scholar used to this sort of argument by contradiction!

The null hypothesis is rarely true

The null hypothesis is that in the populations being studied, the difference between the means equals 0.0000 or the risk ratio equals 1.0000. But scientists rarely conduct experiments or studies where it is even conceivable that the null hypothesis is exactly true. The difference (or correlation, or association) in the overall population might be trivial, but rare is exactly zero. For this reason, the P value answers a question that it rarely makes sense to ask.

Sample size per group	20	100	1000
Incidence in controls	20.0%	20.0%	20.0%
Incidence in treated	10.0%	10.0%	10.0%
Relative risk (RR)	0.50	0.50	0.50
95% CI of RR	0.103 to 2.43	0.247 to 1.00	0.400 to 0.625
P value	0.6614	0.0734	< 0.0001

Table 13.1. Three experiments with the same effect size but different P values. Sample size matters!

P values are not very reproducible

P values are less reproducible than most people would guess. To demonstrate this, I simulated multiple data sets randomly sampled from the same populations. The top portion of Figure 13.1 shows four simulated experiments. The four P values vary considerably. The bottom portion of the figure shows the distribution of P values from 2,500 such simulated experiments. Leaving out the 2.5% highest and lowest P values, the middle 95% of the P values range from 0.0001517 to 0.6869—a range covering more than three orders of magnitude! This result has been confirmed by others (Boos & Stefanski, 2011; Cumming, 2008, 2011). P values are not very reproducible.

LINGO

Note that the variable p or P is used inconsistently. A P value has a very definite meaning as explained in this chapter. But the variable p or P can also be used in a more general sense to mean any probability.

When referring to P values, different books and journals use different styles. Some use lower case p; some upper case P. Some use a hyphen; some do not. The most common style is probably "*p*-value".

COMMON MISTAKES

Many of the points listed below come from Kline (2004) and Goodman (2008).

Mistake: Interpreting the P value as the chance that there is a real difference between populations

Many people misunderstand what a P value means. Let's assume you read that investigators compared a mean in groups of patients given two alternative treatments and reported that the P value equals 0.03. Here are two correct definitions of this P value:

- If the two population means are identical (the null hypothesis is true), there is a 3% chance of observing a difference as large as or larger than observed.
- Random sampling from identical populations will lead to a difference smaller than observed in 97% of experiments and larger than observed in 3% of experiments.

Many people want to believe the definition below, which appears crossed out to emphasize that it is wrong:

~~There is a 97% chance that there is a real difference between populations and a 3% chance that the difference is a random coincidence.~~

Mistake: Thinking that the P value is the probability that the result was due to sampling error

The P value is computed assuming that the null hypothesis is true. In other words, the P value is computed based on the assumption that the difference is due to

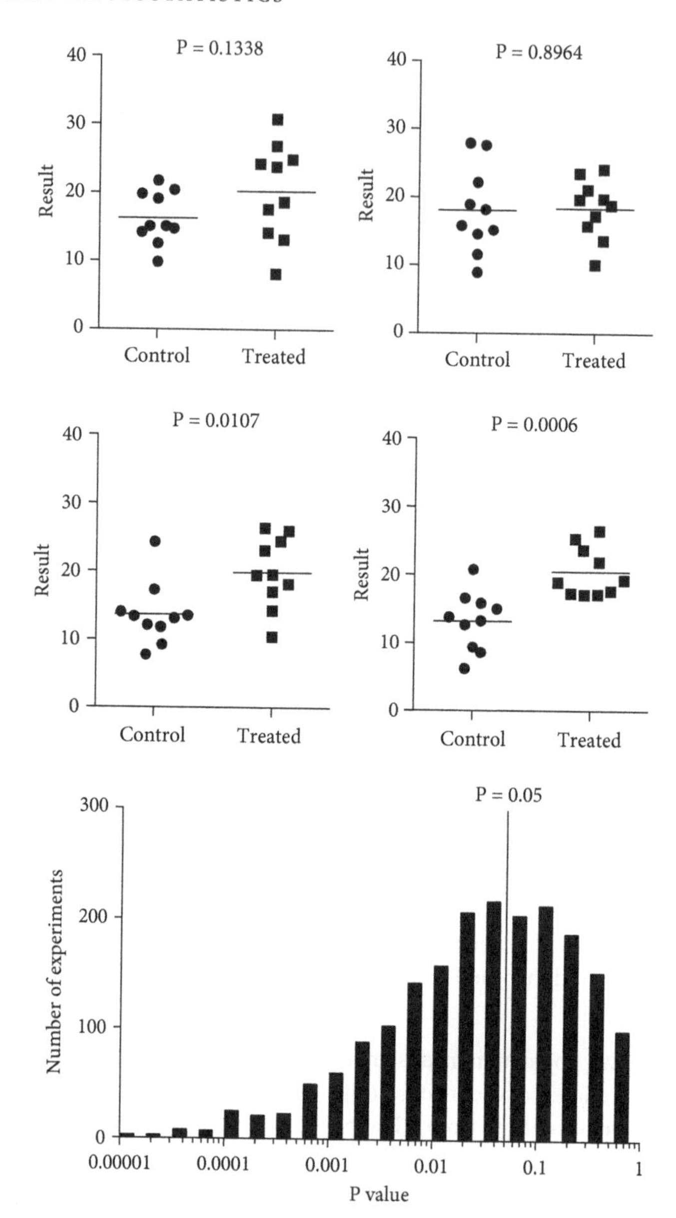

Figure 13.1. P values are not very reproducible.

(Top) Four pairs of data sets simulated (using GraphPad Prism) from Gaussian distributions with a standard deviation equal to 5.0. The population means of the two populations differ by 5.0. Ten values were simulated in each group, and an unpaired t test was used to compare the sample means. Note how far apart the four P values are even though the data in each of the four experiments were sampled from the same populations. (Bottom) Histogram showing the distribution of P values from 2,500 simulated data sets sampled from the same populations. Note that the X axis is logarithmic. The P values span a very wide range.

randomness in selecting subjects—that is, to sampling error. Therefore, the P value cannot tell you the probability that the result is due to sampling error.

Mistake: Interpreting the P value as the probability that the null hypothesis is true

The P value is computed *assuming* that the null hypothesis is true, so it cannot be the probability that it is true.

Mistake: Thinking that the probability that the alternative hypothesis is true is 1.0 minus the P value

If the P value is 0.03, it is very tempting to think that if there is only a 3% probability that a difference would have been caused by random chance, then there must be a 97% probability that it was caused by a real effect. But this is wrong! What you *can* say is that if the null hypothesis were true, then 97% of experiments would lead to a difference smaller than the one observed and 3% of experiments would lead to a difference as large as or larger than the one observed.

Mistake: Thinking that 1.0 minus the P value equals the probability that the results will hold up when the experiment is repeated

If the P value is 0.03, it is tempting to think that this means there is a 97% chance of getting similar results in a repeated experiment. Not so. The P value does not quantify reproducibility.

Mistake: Assuming a large P value proves that the null hypothesis is true

A high P value means that if the null hypothesis were true, it would not be surprising to observe the treatment effect seen in a particular experiment. But that does not prove that the null hypothesis is true. It just means the data are consistent with the null hypothesis, so we don't reject the null hypothesis.

Mistake: Interpreting $P = 0.05$ the same as $P < 0.05$

$P = 0.05$ means what it says—namely, that the P value equals 0.05. In contrast, $P < 0.05$ means that the P value is less than 0.05.

Mistake: Always writing the P value as an inequality

A P value is a number, and it should be expressed as such. When very small, it can make sense to simply say, for example, that $P < 0.0001$. But it is more informative to state the P value as a number (e.g., $P = 0.0345$) than to state some value it is less than (e.g., $P < 0.05$).

Mistake: Focusing only on P values when interpreting data

When interpreting data, look at more than the P value. Also look at the size of the difference (or ratio) and its confidence interval and think about potential biases in the experimental design.

Q & A

Can P values be negative?

No. P values are decimal fractions, so they are always between 0.0 and 1.0.

Can a P value equal 0.0?

A P value can be very small, but it can never equal 0.0. If you see a report that a P value equals 0.0000, that really means it is tiny (perhaps less than 0.001).

Can a P value equal 1.0?

A P value would equal 1.0 only in the rare case when the treatment effect in your sample precisely equals the one defined by the null hypothesis. When a computer program reports that the P value is 1.0000, it usually means that the P value is greater than 0.9999.

Should P values be reported as fractions or percentages?

By tradition, P values are always presented as decimal fractions and never as percentages.

Is a P value always associated with a null hypothesis?

Yes. If you can't state the null hypothesis, then you can't interpret the P value.

Should P values always be presented with a conclusion about whether the results are statistically significant?

No. A P value can be interpreted on its own. In some situations, it can make sense to go one step further and report whether or not the results are statistically significant (as will be explained in Chapter 14). However, this step is optional.

What are one-tailed (or one-sided) P values?

This book only discusses two-tailed (or two-sided) P values. You'll need to read elsewhere to understand one-tailed (or one-sided) P values.

CHAPTER SUMMARY

- P values are frequently reported in scientific papers, so it is essential that every scientist understand exactly what a P value is and is not.
- All P values are based on a null hypothesis. If you cannot state the null hypothesis, you can't understand the P value.
- A P value answers the following general question: If the null hypothesis is true, what is the chance that random sampling will lead to a difference (or association, etc.) as large as or larger than observed in this study?
- P values are calculated values between 0.0 and 1.0. They can be reported and interpreted without ever using the word *significant*.
- If you repeat an experiment, expect the P value to be very different. P values are much less reproducible than you would guess.
- There is much more to statistics than P values. When reading scientific papers, don't get mesmerized by P values. Instead, focus on what the investigators found and the effect size.

CHAPTER 14

Statistical Significance and Hypothesis Testing

STATISTICAL HYPOTHESIS TESTING

With some kinds of analyses, the primary goal is to make a decision. In a pilot experiment of a new drug, the goal may be to decide whether the results are promising enough to merit a second experiment. In a Phase III drug trial, the goal may be to decide whether the drug should be recommended for approval. In quality control, the goal is to decide whether a batch of product can be released.

Statistical hypothesis testing automates decision making. First, define a threshold P value for declaring whether a result is statistically significant. This threshold is called the *significance level* of the test; it is denoted by α (*alpha*) and is commonly set to 0.05. If the P value is less than α, conclude that the difference is *statistically significant* and reject the null hypothesis. Otherwise, conclude that the difference is not statistically significant and do not reject the null hypothesis. That's it.

Note that statistics gives the term *hypothesis testing* a unique meaning very different than what most scientists think of when they go about testing a scientific hypothesis.

REVISITING THE EXAMPLES FROM CHAPTERS 12 AND 13

The conclusions of the five examples in Chapter 12 and 13 can be expressed in terms of statistical significance. For these examples, I set α to its traditional value of 0.05.

- Effect of apixaban on reducing thromboembolism. The P value is less than 0.0001, so the effect is statistically significant.
- Maximum bladder relaxation by norepinephrine in old and young rats. The P value is 0.0003, so this difference is statistically significant.
- Survival of patients with chronic active hepatitis given prednisolone. The P value is 0.031, so the difference in survival between patients given prednisolone and those given placebo is statistically significant.

- Selective serotonin uptake inhibitors and risk of autism. The P value is greater than 0.05, so the association between taking selective serotonin reuptake inhibitors and having a child with autism is not statistically significant.
- Hyperglycemic control in intensive care. The P value is greater than 0.05, so the difference in survival between children in intensive care managed with tight hyperglycemic control and those managed in a standard way is not statistically significant.
- Processed meat and colon cancer. The P value is less than 0.05, so the association between heavy consumption of processed meat and colon cancer is statistically significant.

If your goal is to reduce the results down to a binary decision, then it is helpful to declare a result to be statistically significant or not. But in most cases, I think, it is better to look at the size of the effect and its confidence interval, as was done for these examples in Chapter 12.

ANALOGY: INNOCENT UNTIL PROVEN GUILTY

Scientist as juror

The steps that a jury must follow to determine criminal guilt are very similar to the steps that a scientist follows when using hypothesis testing to determine statistical significance.

A juror starts with the presumption that the defendant is innocent. A scientist starts with the presumption that the null hypothesis of no difference (or no association, or no correlation) is true.

A juror bases his or her decision only on factual evidence presented in the trial, without considering any other information. A scientist bases his or her decision about statistical significance only on data from one experiment, without considering what other experiments have concluded.

A juror reaches the verdict of guilty when the evidence is inconsistent with the assumption of innocence. Otherwise, the juror reaches a verdict of not guilty. A scientist reaches a conclusion that the results are statistically significant when the P value is small enough to make the null hypothesis unlikely. Otherwise, a scientist concludes that the results are not statistically significant.

A juror reaches a verdict of not guilty when guilt has not been proven even if he or she is not convinced that the defendant is innocent. A scientist reaches the conclusion that results are not statistically significant whenever the data are consistent with the null hypothesis even if she or he is not convinced that the null hypothesis is true.

A juror can never reach a verdict that a defendant is innocent. The only choices are guilty or not guilty. A statistical test never concludes that the null hypothesis is true, only that there is insufficient evidence to reject it.

A juror must conclude guilty or not guilty. A juror cannot conclude "I am not sure." Similarly, each statistical test leads to a crisp conclusion than an observed effect is either statistically significant or not statistically significant. A scientist who strictly follows the logic of statistical hypothesis testing cannot conclude "Let's wait for more data before deciding."

Scientist as journalist

Jurors aren't the only people who evaluate evidence presented at a criminal trial. Journalists also evaluate this evidence, but they follow very different rules than jurors. A journalist's job is not to reach a verdict of guilty or not guilty but rather to summarize the proceedings.

Scientists in many fields are often more similar to journalists than jurors. If you don't need to make a clear decision based on one P value, you don't need to use the term *statistically significant* or the rubric of statistical hypothesis testing.

EXTREMELY SIGNIFICANT? BORDERLINE SIGNIFICANT?

Tiny P values

Is a result with P = 0.004 more statistically significant than a result with P = 0.04? By the strict rules of statistical hypothesis testing, the answer is no. Once you have established a significance level, every result is either statistically significant or not statistically significant. Because the goal of statistical hypothesis testing is to make a crisp decision, only two conclusions are needed.

Most scientists are less rigid, however, and refer to *very significant* or *extremely significant* results when the P value is tiny. When showing P values on graphs, investigators commonly use asterisks to create a scale such as that used in *Michelin Guides* or movie reviews (see Table 14.1). When you read this kind of graph, make sure to look at the key that defines the symbols, because different investigators use different threshold values.

Borderline statistical significance

If you follow the strict paradigm of statistical hypothesis testing and set α to its conventional value of 0.05, then a P value of 0.049 denotes a statistically significant difference, and a P value of 0.051 does not. This arbitrary distinction is unavoidable, because the whole point of statistical hypothesis testing is to reach a crisp conclusion from every experiment, without exception.

When a P value is just slightly greater than α, some scientists avoid the phrase "not significantly different" and instead refer to the result as "marginally

SYMBOLS	PHRASE	P VALUE
ns	Not statistically significant	$P > 0.05$
*	Statistically significant	$P < 0.05$
**	Highly significant	$P < 0.01$
***	Extremely significant	$P < 0.001$

Table 14.1. Using asterisks to denote statistical significance.

significant" or "almost statistically significant." Some get more creative and use phrases like "a clear trend," "barely missed statistical significance," "not exactly significant," "not currently significant," "on the cusp of significance," "provisionally significant," "trending toward significance," "verging on significance," "weakly significant," and so on. Hankins (2013) lists 468 such phrases he found in published papers! Rather than deal with linguistic tricks, it is better to just report the actual P value and not worry about whether it is above or below some arbitrary threshold value.

LINGO

Type I, II and III errors

Statistical hypothesis testing makes a decision based on the results of one comparison. When you make this decision, there are several kinds of mistakes that you can make.

These errors are theoretical concepts. When you analyze any particular set of data, you don't know whether the populations are identical. You only know the data in your particular samples. You will never know whether you made one of these errors as part of any particular analysis.

Type I error

When there really is no difference (or association, or correlation) between the populations, random sampling can lead to a difference (or association, or correlation) large enough to be a statistically significant. This is a *Type I error* (see Table 14.2). It occurs when you decide to reject the null hypothesis when in fact the null hypothesis is true. It is a false positive. Chapter 16 will explore Type I errors in more depth.

Type II error

When there really is a difference (or association, or correlation) between the populations, random sampling (and small sample size) can lead to a difference (or association, or correlation) too large to be statistically significant. This is a *Type II error* (see Table 14.2). It occurs when you decide not to reject the null hypothesis when in fact the null hypothesis is false. It is a false negative.

Type III error

What if you conclude that a difference (or association, or correlation) is statistically significant but are wrong about the direction of the effect? You are correct

	DECISION: REJECT NULL HYPOTHESIS	DECISION: DO NOT REJECT NULL HYPOTHESIS
Null hypothesis is true	Type I error	(No error)
Null hypothesis is false	(No error)	Type II error

Table 14.2. Definition of Type I and Type II errors.

that the null hypothesis is wrong, so you haven't made a Type I error. But you've got the direction of the effect backward. In your sample, the new drug works better than the standard drug, but in fact (if you had infinite sample size), it works (on average) worse. Or the data in your sample show that people taking a certain drug lose weight when in fact (if you had an infinite sample) the drug causes people (on average) to gain weight. Gelman and Tuerlinckx (2000) call this a *Type S error*, because the sign of the difference is backward. Hsu (1996) calls this a *Type III error*. Neither term is widely used.

The word "significant"

A result is said to be *statistically significant* when the calculated P value is less than an arbitrary, preset value of α. But a conclusion that a result is statistically significant does not mean the difference is large enough to be interesting, intriguing enough to be worthy of further investigation, or that the finding is scientifically or clinically significant.

To avoid the ambiguity in the word *significant*, never use that word by itself. Always use the phrase "statistically significant" when referring to statistical hypothesis testing and phrases such as "scientifically significant" or "clinically significant" when referring to the size of a difference (or association, or correlation).

Even better, avoid the word *significant* altogether. It is not needed. When writing about statistical hypothesis testing, state the P value and whether or not the null hypothesis is rejected. When discussing the importance or impact of data, banish the word *significant* and instead use words such as *consequential, eventful, meaningful, momentous, large, big, important, substantial, remarkable, valuable, worthwhile, impressive, major*, or *prominent*.

CHOOSING A SIGNIFICANCE LEVEL

The distinction between the P value and α

The P value and α are not the same:

- The significance level (α) is chosen by the experimenter as part of the experimental design before collecting any data. If the null hypothesis is true, α is the probability of rejecting the null hypothesis.
- A P value is computed from the data. You reject the null hypothesis (and conclude that the results are statistically significant) when the P value from a particular experiment is less than the significance level (usually set to 0.05).

The trade-off

A result is deemed statistically significant when the P value is less than α, so you need to choose that significance level. By tradition, α is usually set to 0.05, but you can choose whatever value you want. When choosing a significance level, however, you confront a trade-off.

If you set α to a very low value, you will make few Type I errors. This means that if the null hypothesis is true, there is only a small chance that you will mistakenly call a result statistically significant. However, there is a larger chance that you will not find a significant difference even if the null hypothesis is false. In other words, reducing the value of α will decrease your chance of making a Type I error but increase the chance of a Type II error.

If you set α to a very large value, you will make more Type I errors. This means that if the null hypothesis is true, there is a large chance that you will mistakenly conclude that the effect is statistically significant. However, there is a small chance of missing a real difference. In other words, increasing the value of α will increase your chance of making a Type I error but decrease the chance of a Type II error. The only way to reduce the chances of both a Type I error and a Type II error is to collect larger samples.

Example: Trial by jury

Table 14.3 continues the analogy between statistical significance and the legal system used earlier in this chapter. The relative consequences of Type I and Type II errors depend on the type of trial.

In the United States (and many other countries), a defendant in a criminal trial is considered innocent until proven guilty "beyond a reasonable doubt." This system is based on the principle that it is better to let many guilty people go free than to falsely convict one innocent person. The system is designed to avoid Type I errors in criminal trials—even at the expense of many Type II errors. You could say that α is set to a very low value.

In civil trials, the court or jury rules in favor of the plaintiff if the evidence shows that the plaintiff is "more likely than not" to be right. Here, the thinking is that it is no worse to falsely rule in favor of the plaintiff than to falsely rule in favor of the defendant. The system attempts to equalize the chances of Type I and Type II errors in civil trials.

Example: Detecting spam

Table 14.4 recasts Table 14.2 in the context of detecting spam (junk) email. Spam filters use various criteria to evaluate the probability that a particular email is spam and then use this probability to decide whether to deliver that message to the inbox or the spam folder.

The null hypothesis is that an email is good (not spam). Some spam-filtering software let the user choose whether he or she would rather have more good mail in the spam box (Type I error) or more spam in the inbox (Type II error).

Example: Particle physics and the five-sigma cut-off

In 2012, an international group of physicists announced the discovery of the Higgs boson, because the data met the five-sigma threshold (Lamb, 2012). Meeting the

	VERDICT: GUILTY	VERDICT: NOT GUILTY
Did not commit the crime	Type I error	(No error)
Did commit the crime	(No error)	Type II error

Table 14.3. Type I and Type II errors in the context of trial by jury for a criminal case.

	DECISION: DELETE AS JUNK	DECISION: PLACE IN INBOX
Good email	Type I error	(No error)
Spam	(No error)	Type II error

Table 14.4. Type I and Type II errors in the context of spam filters.

five-sigma threshold means that the results would occur by chance alone as rarely as a value sampled from a Gaussian distribution would be five standard deviations (sigmas) away from the mean. This is equivalent to setting α to 0.0000003. The point here is that there is nothing magic about 0.05, and different fields define α differently.

COMMON MISTAKES

Mistake: Accepting the null hypothesis

If the P value is greater than α, you should conclude that you cannot reject the null hypothesis. You should never say that you accept the null hypothesis. In these cases, the data are consistent with the null hypothesis, but that does not prove the null hypothesis is true.

Mistake: Stargazing

Many scientists mistakenly believe the only thing that matters is whether a result is statistically significant, a belief that has even been called a cult (Ziliak & McCloskey, 2008). Because statistical significance is often shown on graphs and tables as asterisks or stars, the process of overemphasizing statistical significance is sometimes referred to as *stargazing*. Don't focus only on whether a result is statistically significant. Look at the size of the effect and its precision as quantified by the confidence interval.

Mistake: Believing that if a result is statistically significant, the effect must be large

A conclusion that a finding is statistically significant:

- Does *not* mean the difference is large enough to be interesting. If the sample size is large, then even tiny, inconsequential differences will be statistically significant. We'll return to this point in the next chapter.
- Does *not* mean the results are intriguing enough to be worthy of further investigation.
- Does *not* mean that the finding is scientifically or clinically significant.

Mistake: P-hacking to obtain statistical significance

One problem with statistical hypothesis testing is that investigators may be tempted to work hard to push their P value low enough to be declared significant. They can do this in a number of ways:

- Tweaking. The investigators may have played with the analyses. If one analysis didn't give a P value less than 0.05, then they tried a different one. Perhaps they switched between parametric and nonparametric analysis or tried including different variables in multiple regression models. Or perhaps they reported only the analyses that gave P values less than 0.05 and ignored the rest.

- Using dynamic sample size. The investigators may have analyzed their data several times. And each time, they may have stopped if the P value was less than 0.05 but collected more data when the P value was above 0.05. This approach would yield misleading results, because it is biased toward stopping when P values are small.
- Analyzing subgroups separately. The investigators may have analyzed various subsets of the data and only reported the subsets that gave low P values or failed to adjust for multiple comparisons (see Chapter 17).
- Reporting results of multiple outcomes selectively. If several outcomes were measured, the investigators may have chosen to report only those for which the P value was less than 0.05.
- Playing with outliers. The investigators may have tried using various definitions of outliers, reanalyzed the data several times, and only reported the analyses with low P values.

Simmons, Nelson, and Simonsohn (2011) coined the term *P-hacking* to refer to attempts by investigators to lower the P value by trying various analyses or by analyzing subsets of data.

Q & A

Is it possible to report scientific data without using statistical hypothesis testing?

Yes. Report the data along with the confidence intervals and, perhaps, the P values.

It is possible to apply statistical hypothesis testing without using the word *significant*?

Sure! The term *significant* is often misunderstood, so it makes sense to avoid this word. If you want to apply the ideas of statistical hypothesis testing, state the null hypothesis and whether or not you reject it. The term *significant* is not necessary.

But isn't the whole point of statistics to decide when an effect is statistically significant?

No. The goal of statistics is to quantify scientific evidence and uncertainty.

Are the P value and α the same?

No. A P value is computed from the data. The significance level, or α, is chosen by the experimenter, as part of the experimental design, before collecting any data. A difference is termed *statistically significant* if the P value computed from the data is less than the value of α set in advance.

If the 95% CI just barely reaches the value that defines the null hypothesis, what can you conclude about the P value?

If the 95% CI includes the value that defines the null hypothesis, you can conclude that the P value is greater than 0.05. If the 95% CI excludes the value that defines the null hypothesis, you can conclude that the P value is less than 0.05. So if the 95% CI ends right at the value that defines the null hypothesis, then the P value must be very close to 0.05.

If the 95% CI is centered on the value that defines the null hypothesis, what can you conclude about the P value?

The observed outcome must equal the value that defines the null hypothesis, so the P value must equal 1.000.

CHAPTER SUMMARY

- Statistical hypothesis testing reduces all findings to two conclusions, statistically significant or not statistically significant.
- The distinction between the two conclusions is based purely on whether the computed P value is less than or not less than an arbitrary threshold, called the *significance level*.
- A conclusion of statistical significance does not mean that the difference is large enough to be interesting, that the results are intriguing enough to be worthy of further investigations, or that the finding is scientifically or clinically significant.
- The concept of statistical significance is often overemphasized. The whole idea of statistical hypothesis testing is only useful when a definitive decision needs to be made based on one analysis.
- If the 95% CI includes the null hypothesis, then the P value must be less than 0.05. Conversely, if the 95% CI does not include the null hypothesis, then the P value must be less than 0.05.
- The term *significant* can be confusing, because it can refer to the fact that a P value is low enough to reject the null hypothesis and to an effect size (or difference, or association) that is large enough to be scientifically or clinically important.
- Some scientists use the phrase "very significant" or "extremely significant" when a P value is tiny and "borderline significant" when a P value is just a bit greater than 0.05.
- While it is conventional to use 0.05 as the threshold P value, or α, that separates statistically significant from not statistically significant, this value is totally arbitrary.
- Ideally, you should choose α based on the consequences of Type I (false-positive) and Type II (false-negative) errors.
- Distinguish the P value from α. The P value is computed from data, while the value of α is (or should be) decided as part of the experimental design.

CHAPTER 15

Interpreting a Result That Is (Or Is Not) Statistically Significant

INTERPRETING RESULTS THAT ARE "STATISTICALLY SIGNIFICANT"

Let's consider a simple scenario: comparing enzyme activity in cells incubated with a new drug to enzyme activity in control cells. Your scientific hypothesis is that the drug increases enzyme activity, so the null hypothesis is that there is no difference. You run the analysis, and indeed, the enzyme activity is higher in the treated cells. You run a t test and find that the P value is less than 0.05, so you conclude the result is statistically significant. Below are seven possible explanations for why this happened.

Explanation 1: The drug had a substantial effect

Scenario: The drug really does have a substantial effect to increase the activity of the enzyme you are measuring.

Discussion: This is what everyone thinks when they see a small P value in this situation. But it is only one of seven possible explanations.

Explanation 2: The drug had a trivial effect

Scenario: The drug actually has a very small effect on the enzyme you are studying even though your experiment yielded a P value of less than 0.05.

Discussion: The small P value just tells you that the difference is large enough to be surprising if the drug really didn't affect the enzyme at all. It does not imply that the drug has a large treatment effect. A small P value with a small effect will happen with some combination of a large sample size and low variability. It will also happen if the treated group has measurements a bit above their true average and the control group happens to have measurements a bit below their true average, causing the effect observed in your experiment to be larger than the true effect.

Explanation 3: There was a Type I error or false discovery

Scenario: The drug really doesn't affect enzyme expression at all. Random sampling just happened to give you higher values in the cells treated with the drug and lower levels in the control cells. Accordingly, the P value was small.

Discussion: This is called a *Type I error* or *false discovery*. If the drug really has no effect on the enzyme activity (if the null hypothesis is true) and you choose the traditional 0.05 significance level ($\alpha = 5\%$), you'll expect to make a Type I error 5% of the time. Therefore, it is easy to think that you'll make a Type I error 5% of the time when you conclude that a difference is statistically significant. However, this is not correct. You'll understand why when you read about the false discovery rate (the fraction of "statistically significant" conclusions that are wrong) in the next chapter.

Explanation 4: There was a Type S error

Scenario: The drug (on average) decreases enzyme activity, which you would only know if you repeated the experiment many times. In this particular experiment, however, random sampling happened to give you high enzyme activity in the drug-treated cells and low activity in the control cells, and this difference was large enough, and consistent enough, to be statistically significant.

Discussion: Your conclusion is backward. You concluded that the drug (on average) significantly increases enzyme activity when in fact the drug decreases enzyme activity. You've correctly rejected the null hypothesis that the drug does not influence enzyme activity, so you have not made a Type I error. Instead, you have made a *Type S error*, because the *sign* (plus or minus, increase or decrease) of the actual overall effect is opposite to the results you happened to observe in one experiment. Type S errors are rare but not impossible. They are also referred to as *Type III errors*.

Explanation 5: The experimental design was poor

Scenario: The enzyme activity really did go up in the tubes with the added drug. However, the reason for the increase in enzyme activity had nothing to do with the drug but rather with the fact that the drug was dissolved in an acid (and the cells were poorly buffered) and the controls did not receive the acid (due to bad experimental design). So the increase in enzyme activity was actually due to acidifying the cells and had nothing to do with the drug itself. The statistical conclusion was correct—adding the drug did increase the enzyme activity—but the scientific conclusion was completely wrong.

Discussion: Statistical analyses are only a small part of good science. That is why it is so important to design experiments well, to randomize and blind when possible, to include necessary positive and negative controls, and to validate all methods.

Explanation 6: The results cannot be interpreted due to ad hoc multiple comparisons

Scenario: You actually ran this experiment many times, each time testing a different drug. The results were not statistically significant for the first 24 drugs tested but the results with the 25th drug were statistically significant.

Discussion: These results would be impossible to interpret, as you'd expect some small P values just by chance when you do many comparisons. If you know how many comparisons were made (or planned), you can correct for multiple comparisons. But here, the design was not planned, so a rigorous interpretation is impossible. We'll return to this topic in Chapter 17.

Explanation 7: The results cannot be interpreted due to dynamic sample size

Scenario: You first ran the experiment in triplicate, and the result (n = 3) was not statistically significant. Then you ran it again and pooled the results (so now n = 6), and the results were still not statistically significant. So you ran it again, and finally, the pooled results (with n = 9) were statistically significant.

Discussion: The P value you obtain from this approach simply cannot be interpreted. P values can only be interpreted at face value when the sample size, the experimental protocol, and all the data manipulations and analyses were planned in advance. Otherwise, you are P-hacking, and the results cannot be interpreted. We'll return to this topic in Chapter 18.

Bottom line: When a P value is small, consider all the possibilities

Be cautious when interpreting small P values. Don't make the mistake of instantly believing explanation 1 above without also considering the possibility that the true explanation is one of the other six possibilities listed. Read on to see how you need to be just as wary when interpreting a high P value.

INTERPRETING RESULTS THAT ARE "NOT STATISTICALLY SIGNIFICANT"

"Not significantly different" does not mean "no difference"

A large P value means that a difference (or correlation, or association) as large as what you observed would happen frequently as a result of random sampling. But this does not necessarily mean that the null hypothesis of no difference is true or that the difference you observed is definitely the result of random sampling.

Vickers (2010) tells a great story that illustrates this point:

> The other day I shot baskets with [the famous basketball player] Michael Jordan (remember that I am a statistician and never make things up). He made 7 straight free throws; I made 3 and missed 4 and then (being a statistician) rushed to the sideline, grabbed my laptop, and calculated a P value of .07 by Fisher's exact test. Now, you wouldn't take this P value to suggest that there is *no* difference between my basketball skills and those of Michael Jordan, you'd say that our experiment hadn't *proved* a difference.

A high P value does not prove the null hypothesis. And deciding not to reject the null hypothesis is not the same as believing that the null hypothesis

	CONTROLS	HYPERTENSION
Number of subjects	17	18
Mean receptor number (receptors/platelet)	263	257
Standard deviation	87	59

Table 15.1. Number of α_2-adrenergic receptors on the platelets isolated from the blood of controls and people with hypertension.

is definitely true. The absence of evidence is not evidence of absence (Altman & Bland, 1995).

Example: α_2-Adrenergic receptors on platelets

Epinephrine, acting through α_2-adrenergic receptors, makes blood platelets stickier and thus helps blood clot. Motulsky, O'Connor, and Insel (1983) counted these receptors and compared people with normal and high blood pressure. The idea was that the adrenergic signaling system might be abnormal in people with high blood pressure (hypertension). We were most interested in the effects on the heart, blood vessels, kidney, and brain but obviously couldn't access those tissues in people, so we counted the receptors on platelets instead. Table 15.1 shows the results.

The results were analyzed using a test called the *unpaired t test*, also called the *two-sample t test*. The average number of receptors per platelet was almost the same in both groups, so of course, the P value was high—specifically, 0.81. If the two populations were Gaussian with identical means, you'd expect to see a difference as large as or larger than that observed in this study in 81% of studies of this size.

Clearly, these data provide no evidence that the mean receptor number differs in the two groups. When my colleagues and I published this study over 30 years ago, we stated that the results were not statistically significant and stopped there, implying that the high P value proves that the null hypothesis is true. But that was not a complete way to present the data. We should have interpreted the confidence interval.

The 95% CI for the difference between group means extends from –45 to 57 receptors per platelet. To put this in perspective, the average number of α_2-adrenergic receptors per platelet is about 260. We can therefore rewrite the confidence interval as extending from –45/260 to 57/260, which is from –17.3% to 21.9%, or approximately plus or minus 20%.

It is only possible to properly interpret the confidence interval in a scientific context. Here are two alternative ways to think about these results:

- A 20% change in receptor number could have a huge physiological impact. With such a wide confidence interval, the data are inconclusive, because you could have these data with no difference, substantially more receptors, or substantially fewer receptors on platelets of people with hypertension.

- The confidence interval convincingly shows that the true difference is unlikely to be more than 20% in either direction. This experiment counts receptors on a convenient tissue (blood cells) as a marker for other organs, and we know that the number of receptors per platelet varies a lot from individual to individual. For these reasons, we'd only be intrigued by the results (and want to pursue this line of research) if the receptor number in the two groups differed by at least 50%. Here, the 95% CI extended about 20% in each direction. Therefore, we can reach a solid negative conclusion that the receptor number in individuals with hypertension does not change or that any such change is physiologically trivial and not worth pursuing.

The difference between these two perspectives is a matter of scientific judgment. Would a difference of 20% in receptor number be scientifically relevant? The answer depends on scientific (physiological) thinking. Statistical calculations have nothing to do with it. Statistical calculations are only a small part of interpreting data.

FIVE EXPLANATIONS FOR "NOT STATISTICALLY SIGNIFICANT" RESULTS

Let's continue the simple scenario from the first part of this chapter. You compare cells incubated with a new drug to control cells and measure the activity of an enzyme, and you find that the P value is large enough (greater than 0.05) for you to conclude that the result is not statistically significant. Below are five explanations to explain why this happened.

Explanation 1: The drug did not affect the enzyme you are studying

Scenario: The drug did not induce or activate the enzyme you are studying, so the enzyme's activity is the same (on average) in treated and control cells.

Discussion: This is, of course, the conclusion everyone jumps to when they see the phrase "not statistically significant." However, four other explanations are possible.

Explanation 2: The drug had a trivial effect

Scenario: The drug may actually affect the enzyme but by only a small amount.

Discussion: This explanation is often forgotten.

Explanation 3: There was a Type II error

Scenario: The drug really did substantially affect enzyme expression. Random sampling just happened to give you some low values in the cells treated with the drug and some high levels in the control cells. Accordingly, the P value was large, and you conclude that the result is not statistically significant.

Discussion: How likely are you to make this kind of Type II error? It depends on how large the actual (or hypothesized) difference is, on the sample size, and on the experimental variation. We'll return to this topic in Chapter 18.

Explanation 4: The experimental design was poor

Scenario: In this scenario, the drug really would increase the activity of the enzyme you are measuring. However, the drug was inactivated, because it was dissolved in an acid. Since the cells were never exposed to the active drug, of course the enzyme activity didn't change.

Discussion: The statistical conclusion was correct—adding the drug did not increase the enzyme activity—but the scientific conclusion was completely wrong.

Explanation 5: The results cannot be interpreted due to dynamic sample size

Scenario: In this scenario, you hypothesized that the drug would not work, and you really want the experiment to validate your prediction (maybe you have made a bet on the outcome). You first ran the experiment three times, and the result (n = 3) was statistically significant. Then you ran it three more times, and the pooled results (now with n = 6) were still statistically significant. Then you ran it four more times, and finally, the results (with n = 10) were not statistically significant. This n = 10 result (not statistically significant) is the one you present.

Discussion: The P value you obtain from this approach simply cannot be interpreted.

LINGO

As already mention in Chapter 14, beware of the word *significant.* Its use in statistics as part of the phrase *statistically significant* is very different than the ordinary use of the term to mean important or notable. Whenever you see the word significant, make sure you understand the context.

If you are writing scientific papers, the best way to avoid the ambiguity is to never use the word *significant.* You can write about whether a P value is greater or less than a preset threshold without using that word. And there are lots of alternative words to use to describe the clinical, physiological or scientific impact of a finding.

COMMON MISTAKES

Mistake: Believing that a "statistically significant" result proves an effect is real

As this chapter points out, there are several reasons why a result can be statistically significant.

Mistake: Believing that a "not statistically significant" result proves no effect is present

As this chapter points out, there are several reasons why a result can be not statistically significant.

Mistake: Believing that if a difference is "statistically significant," it must have a large physiological or clinical impact

If the sample size is large, a difference can be statistically significant but also so tiny that it is physiologically or clinically trivial.

Q & A

Is this chapter trying to tell me that it isn't enough to determine if a result is, or is not, statistically significant. Instead, I actually have to think?

Yes!

CHAPTER SUMMARY

- A conclusion of statistical significance can arise in seven different situations:
  - There really is a substantial effect. This is what everyone thinks of when they hear "statistically significant."
  - There is a trivial effect.
  - There is a Type I error or false discovery. In other words, there truly is zero effect, but random sampling caused an apparent effect.
  - There is a Type S error. In other words, the effect is real, but random sampling led you to see an effect in this one experiment that is opposite in direction to the true effect.
  - The study design is poor.
  - The results cannot be interpreted due to ad hoc multiple comparisons.
  - The results cannot be interpreted due to ad hoc sample size selection.
- The word "significant" is often misinterpreted it. Avoid using it. When reading papers that include the word "significant" make sure you know what the author meant.
- "Not significantly different" does not mean "no difference." Absence of evidence is not evidence of absence.
- A conclusion of "not statistically significant" can arise in five different situations:
  - There is no effect. In other words, the null hypothesis is true.
  - The effect is too small to detect in this study.
  - There is a Type II error. In other words, random sampling prevented you from seeing a true effect.
  - The study design is poor.
  - The results cannot be interpreted due to ad hoc sample size selection.

CHAPTER 16

How Common Are Type I Errors?

WHAT IS A TYPE I ERROR?

Type I errors (defined in Chapter 14) are also called *false discoveries* or *false positives*. You run an experiment, do an analysis, find that the P value is less than 0.05 (or whatever threshold you choose), and so reject the null hypothesis and conclude that the difference is statistically significant. You've made a Type I error when the null hypothesis is really true and the tiny P value was simply due to random sampling. If comparing two means, for example, random sampling may result in larger values in one group and smaller values in the other, with a difference large enough (compared to the variability and accounting for sample size) to be statistically significant.

HOW FREQUENTLY DO TYPE I ERRORS OCCUR?

There are two ways to quantify how often Type I errors occur: the significance level (α) and the false discovery rate.

The significance level (α)

The *significance level* (defined in Chapter 14) answers these equivalent questions:

- If the null hypothesis is true, what is the probability that a particular experiment will collect data that generate a P value low enough to reject that null hypothesis?
- Of all experiments you could conduct when the null hypothesis is actually true, in what fraction will you reach a conclusion that the results are statistically significant?

The false discovery rate

The *false discovery rate* (abbreviated FDR) is the answer to these equivalent questions:

- If a result is statistically significant, what is the probability that the null hypothesis is really true?
- Of all experiments that reach a statistically significant conclusion, what fraction are false positives (Type I errors)?

	DECISION: REJECT NULL HYPOTHESIS	DECISION: DO NOT REJECT NULL HYPOTHESIS	TOTAL
Null hypothesis is true	A (Type I error)	B	A + B
Null hypothesis is false	C	D (Type II error)	C + D
Total	A + C	B + D	A + B + C + D

Table 16.1. The results of many hypothetical statistical analyses to reach a decision to reject or not reject the null hypothesis.

In this table, A, B, C, and D are integers (not proportions) that count numbers of analyses (number of P values). The total number of analyses equals A + B + C + D. The significance level is defined to equal A/(A + B). The false discovery rate is defined to equal A/(A + C). The significance level and the false discovery rate are answers to different questions and so rarely are equal.

The false discovery rate and the significance level are not the same

The significance level and the false discovery rate are the answers to distinct questions, so the two are defined differently and their values are usually different. To understand this conclusion, let's do a bit of math. Table 16.1 shows the results of many hypothetical statistical analyses that each reach a decision to reject or not reject the null hypothesis. The top row tabulates results for experiments in which the null hypothesis is really true. The second row tabulates results for experiments in which the null hypothesis is not true. When you analyze data, you don't know whether the null hypothesis is true, so you could never actually create this table from an actual series of experiments.

The false discovery rate only considers analyses that reject the null hypothesis so only deals with the left column of the table. Of all these experiments (A + C), the number in which the null hypothesis is actually true equals A. So the false discovery rate equals A/(A + C).

The significance level only considers analyses where the null hypothesis is true so only deals with the top row of the table. Of all these experiments (A + B), the number of times the null hypothesis is rejected equals A. So the significance level is expected to equal A/(A + B).

THE PRIOR PROBABILITY INFLUENCES THE FALSE DISCOVERY RATE (A BIT OF BAYES)

What is the false discovery rate? Its value depends, in part, on the significance level you choose. But it also depends, in part, upon the context of the experiment. This is a really important point that will be demonstrated via four examples summarized in Table 16.2.

Example 1: Prior probability = 0%

In a randomized clinical trial, each subject is randomly assigned to receive one of two (or more) treatments. Before any treatment is given, clinical investigators commonly compare variables such as the hematocrit (blood count), fasting glucose (blood sugar), blood pressure, weight, and so on.

	PRIOR PROBABILITY	FDR DEFINING "DISCOVERY" AS P < 0.05	FDR DEFINING "DISCOVERY" AS P BETWEEN 0.045 AND 0.050
Example 1: Comparing randomly assigned groups in a clinical trial before treatment	0%	100%	100%
Example 2: Testing a drug that might work	10%	36%	78%
Example 3: Testing a drug that is likely to work	50%	5.9%	27%
Example 4: Positive controls	100%	0%	0%

Table 16.2. The false discovery rate depends on the prior probability and P value. These examples are discussed in the text. Examples 2 and 3 assume an experiment with sufficient sample size to have a power of 80%. All examples assume that the significance level was set to the traditional 5%. The false discovery rate is computed defining "discovery" as all results where the P value is less than 0.05 and as only those results where the P value is between 0.045 and 0.050.

Since these measurements are made before any treatments are given, you can be absolutely, positively sure that both samples are drawn from the same population. In other words, you know for sure that all the null hypotheses are true. The *prior probability* (the probability before collecting data or running a statistical test) that there is a true difference between populations equals zero. If one of the comparisons results in a small P value, you know for sure that you have made a Type I error. In Table 16.1, C and D must equal zero, so A/(A + C), the false discovery rate, equals 1.0, or 100%.

Does it even make sense to compare relevant variables at the beginning of a clinical trial? Sure. If a clinically important variable differs substantially between the two randomly chosen groups before any treatment or intervention, it will be impossible to interpret the study results. You won't be able to tell if any difference in the outcome variable is due to treatment or to unlucky randomization giving you very different kinds of people in the two groups. So it makes sense to compare the groups in a randomized trial before treatment and ask whether the differences in relevant variables prior to treatment are large enough to matter, defining "large enough" either physiologically or clinically (but not based on a P value).

Example 2: Prior probability = 10%

Imagine that you work at a drug company and are screening drugs as possible treatments for hypertension. You are interested in a mean decrease in blood pressure of 10 mmHg or more, and your samples are large enough that there is an 80% chance of finding a statistically significant difference (P < 0.05) if the true difference between population means is at least 10 mmHg. (You will learn how to calculate the sample size in Chapter 18.)

You test a drug that is known to weakly block angiotensin receptors, but the affinity is low and the drug is unstable. From your experience with such drugs,

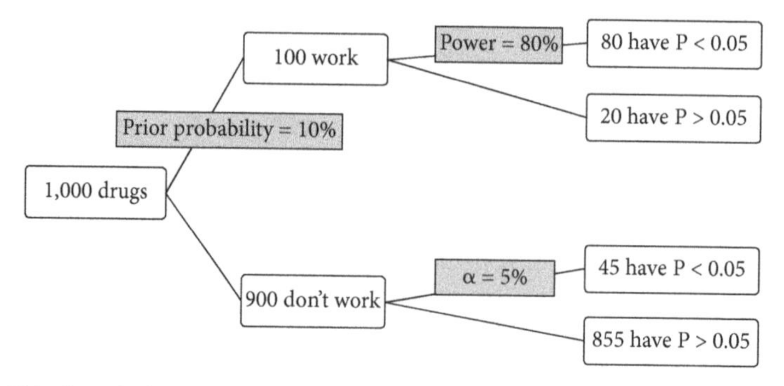

Figure 16.1. A probability tree of example 2.
The FDR equals 45/(80 + 45), or 36%.

you estimate about a 10% chance that it will depress blood pressure. In other words, the prior probability that the drug works is 10%. What can we expect to happen if you test 1,000 such drugs? Figure 16.1 shows the probability tree.

- Of the 1,000 drugs screened, we expect that 100 (10%) will really work.
- Of the 100 drugs that really work, we expect to obtain a statistically significant result in 80 (because our experimental design has 80% power).
- Of the 900 drugs that are really ineffective, we expect to obtain a statistically significant result in 5% (because we set α equal to 0.05). In other words, we expect 5% × 900, or 45, false positives.
- Of 1,000 tests of different drugs, we therefore expect to obtain a statistically significant difference in 80 + 45 = 125. The false discovery rate equals 45/125 = 36%.

Example 3: Prior probability = 50%

In this example, the drug is much better characterized. The drug blocks the right kinds of receptors with reasonable affinity, and it is chemically stable. From your experience with such drugs, you estimate the prior probability that the drug is effective equals 50%. What can we expect to happen if you test 1,000 such drugs? Figure 16.2 shows the probability tree.

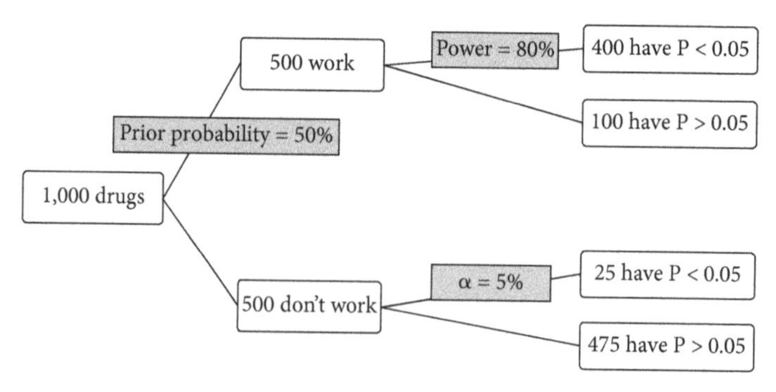

Figure 16.2. A probability tree of example 3.
The FDR equals 25/(400 + 25), or 5.9%.

- Of those 1,000 drugs screened, we expect that 500 (50%) will really work.
- Of the 500 drugs that really work, we expect to obtain a statistically significant result in 400 (because our experimental design has 80% power).
- Of the 500 drugs that are really ineffective, we expect to obtain a statistically significant result in 5% (because we set α equal to 0.05). In other words, we expect 5% × 500, or 25, false positives.
- Of 1,000 tests of different drugs, therefore, we expect to obtain a statistically significant difference in 400 + 25 = 425. The false discovery rate equals 25/425 = 5.9%

Example 4: Prior probability = 100%

You are running a screening assay to find a new drug that works on a particular pathway. You screen tens of thousands of compounds, and none (so far) work. To prove to yourself (and others) that your assay really will detect an active drug, you run a positive control every day that tests a drug already known to work on that pathway. This is important, because you want to make sure that all the negative results you obtain are the result of screening inactive drugs, not because the experimental system is simply not working on a given day (maybe you left out one ingredient).

The drug in the positive control always provokes a big response, so the P value is always less than 0.05. This means that when you run the positive control 100 times, you get a statistically significant result 100 times, without a single false positive. In fact, a false positive (Type I error) is impossible, because you are repeatedly testing a drug that is known to work. So the false discovery rate is 0%.

Bayes

The logic and calculations explained above combine the prior probability with the results of an experiment. This is called a *Bayesian approach*, named after Thomas Bayes, who first published work on this problem in the mid-18th century. This book contains little about Bayesian statistics, but it is a well-developed field of statistical thinking widely used in some scientific fields.

ANALOGY TO CLINICAL TESTING

The DoubleCheckGold™ HIV 1&2 test rapidly detects antibodies to human immunodeficiency virus (abbreviated HIV), the cause of AIDS, in human serum or plasma (Alere, 2014). Its sensitivity is 99.9%, which means that 99.9% of people who are infected with HIV will test positive. Its specificity is 99.6%, which means that 99.6% of people without the infection will test negative, which leaves 100.0% − 99.6% = 0.4% of them who will test positive.

What happens if we test a million people in a population where the prevalence of HIV infection is 10%? The number of individuals with HIV will be 100,000 (10% of 1 million), and 99,900 of those will test positive (99.9% of 100,000). The number of people without HIV will be 900,000. Of those, 0.4% will test positive, which is 3,600 people. In total, there will be 99,900 + 3,600 = 103,500 positive tests. Of these, 3,600/103,500 = 0.035 = 3.5% will be false positives.

What if the prevalence of HIV is 0.1%? The number of individuals with HIV will be 1,000 (0.1% of 1 million), and 999 of those will test positive (99.9% of 1,000). The number of people without HIV will be 999,000. Of those, 0.4% will test positive, which is 3,996 people. In total, there will be 999 + 3,996 = 4,995 positive tests. Of these, 3,996/4,995 = 0.80 = 80% will be false positives.

These examples demonstrate that the fraction of the positive tests that are false positives depends on the prevalence of the disease in the population you are testing. Note the analogy to the false discovery rate for statistical tests, which depends on the prior probability of the hypothesis being tested. The false discovery rate depends on both the context of the test (prior probability for statistical tests, prevalence for clinical tests) and the accuracy of the test (α and power for statistical tests, sensitivity and specificity for clinical tests).

LINGO

The term *false discovery* is used to describe an incorrect conclusion due to random sampling error. But keep in mind that even if an effect is real (not due to random sampling) you may misunderstand the reason for the effect. One example was given in Chapter 15, where the enzyme activity increased in the tubes to which you added drug but the reason for the increase was not the drug itself but rather the fact that the drug had been dissolved in an acid. Due to really poor experimental design, the cells were poorly buffered, and the controls did not receive the acid. So the increase in enzyme activity was actually due to acidifying the cells and not the drug. The statistical conclusion was correct—adding the drug did increase the enzyme activity—but the scientific conclusion was false.

Don't mix up statistical thinking, which can help you decide whether an effect is real or due to chance, with scientific thinking, which requires that you think of alternative explanations for real results.

COMMON MISTAKES

Mistake: Asking about the chance of a Type I error without any clarifying details

How often will you make Type I errors? This chapter has pointed out that are two answers because there really are two questions.

- If the null hypothesis is true, what is the chance of making a Type I error?
- If a finding is statistically significant, what is the chance you made a Type I error?

Mistake: Thinking that the false discovery rate equals the significance level

The false discovery rate is not the same as the significance level. Mixing up the two is a common mistake. Remember, α is the fraction of experiments performed that, when the null hypothesis is true, result in a conclusion that the effect is statistically significant. The false discovery rate is the fraction of experiments leading to the conclusion that the effect is statistically significant where in fact the null hypothesis is really true.

Mistake: Not realizing that the false discovery rate depends on the scientific context

The false discovery rate depends upon the scientific context, as quantified by the prior probability.

Q & A

Don't experienced scientists often account for the prior probability informally, without computing a false discovery rate?

Yes. Here are three examples:

- This study tested a hypothesis that is biologically sound and supported by previous data. The P value is 0.04. I have a choice of believing that the results occurred by a coincidence that will happen 1 time in 25 under the null hypothesis or that the experimental hypothesis is true. Because the experimental hypothesis makes so much sense, I'll believe it. The null hypothesis is probably false.
- This study tested an experimental hypothesis that makes no biological sense and has not been supported by any previous data. The P value is again 0.04. I have a choice of believing that the results occurred by a coincidence that will happen 1 time in 25 under the null hypothesis or that the experimental hypothesis is true. Because the experimental hypothesis is so unlikely to be correct, I think that the results are the result of coincidence. The null hypothesis is probably true.
- This study tested the same experimental hypothesis that makes no biological sense and has not been supported by any previous data. The sample size is much larger, however, and the P value is incredibly low (0.000001). There are no biases or flaws in the experimental design, and the data are reported honestly. I have a choice of believing that the results occurred by a coincidence that will happen 1 time in 1 million under the null hypothesis or that the experimental hypothesis is true. Although that hypothesis seems crazy, the data force me to believe it. The null hypothesis is probably false.

It is appropriate to interpret experimental data in the context of scientific theory and previous data, so it makes sense that different people legitimately reach different conclusions from the same data.

If a study has a P value just a tiny bit less than 0.05, doesn't it make sense to compute the FDR just for P values a tiny bit less than 0.05 and not for all P values less than 0.05?

The right column of Table 16.2 shows the FDR determined this way (based on Colquohoun, 2014). The FDR, of course, is much higher when you consider only P values close to 0.05.

CHAPTER SUMMARY

- You've made a Type I error (a false discovery) when your analysis leads you to conclude that an observed difference or effect is statistically significant when in fact it is an artifact of random sampling.
- The question "What is the chance of making a false discovery?" is ambiguous.

- The significance level (α) answers the question "If the null hypothesis is true, what is the probability of incorrectly rejecting it?"
- The false discovery rate answers the question "Of all experiments that reach a statistically significant conclusion, what fraction are false positives?"
- The false discovery rate depends on the context of the study. In other words, it depends on the prior probability that your hypothesis is true (based on prior data and theory).
- Even if you don't do formal Bayesian calculations, you should consider prior knowledge and theory when interpreting data.

CHAPTER 17

Multiple Comparisons

WHY MULTIPLE COMPARISONS ARE A PROBLEM

If you make two independent comparisons, what is the probability that one or both comparisons will result in a statistically significant conclusion just by chance? Actually, it is easier to answer the opposite question. Assuming that both null hypotheses are true, what is the chance that both comparisons will *not* be statistically significant? The answer is the chance that the first comparison will not be statistically significant (0.95) times the chance that the second comparison will not be statistically significant (also 0.95), which equals 0.9025, or about 90%. That leaves about a 10% probability of obtaining at least one statistically significant conclusion by chance. Figure 17.1 plots this probability for various numbers of independent comparisons.

Consider the unlucky number 13. If you perform 13 independent comparisons (with the null hypothesis true in all cases), the chance is about 50% that one or more of these P values will be less than 0.05 and thus lead to a conclusion that the effect is statistically significant. With more than 13 comparisons, it is more likely than not that one or more conclusions will be statistically significant just by chance. With 100 independent null hypotheses that are all true, the chance of obtaining at least one statistically significant P value is 99%. With *multiple comparisons*, it is easy to be mislead by small P values.

A DRAMATIC DEMONSTRATION OF THE PROBLEM WITH MULTIPLE COMPARISONS

Bennett and colleagues (2011) dramatically demonstrated the problem of multiple comparisons. They used functional magnetic resonance imaging to map blood flow in thousands of areas of a brain. Their experimental subject was shown a photograph of a person and asked to identify which emotion the person in the photograph was experiencing. The investigators measured blood flow to thousands of areas of the brain before and after presenting the photograph to their experimental subject. In two particular areas of the brain, the measure of blood flow increased substantially, and those differences were statistically significant ($P < 0.001$).

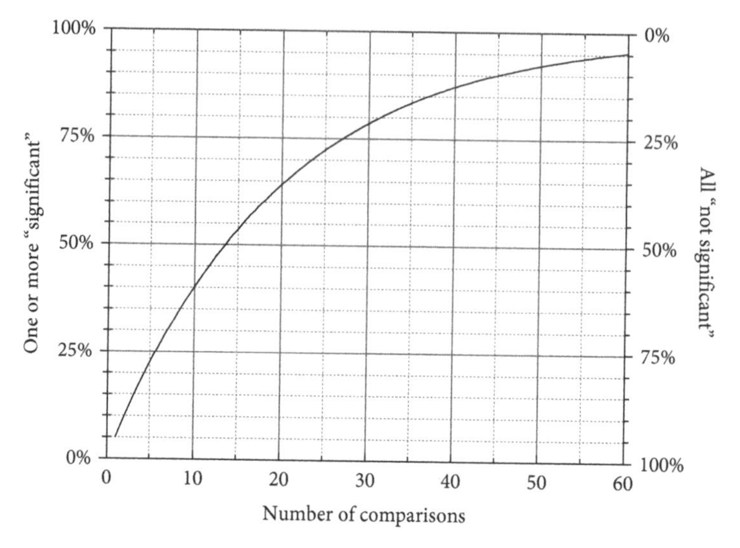

Figure 17.1. Probability of obtaining statistically significant results by chance. The X axis shows various numbers of statistical comparisons, each assumed to be independent of the others. The left Y axis shows the chance of obtaining one or more statistically significant results ($P < 0.05$) by chance. The value on the left Y axis is computed as $1.0 - 0.95^X$, where X is the number of comparisons. The right Y axis shows the chance that all the comparisons will be not statistically significant (equal to 100% minus the value on the left Y axis).

Sounds compelling. The investigators have identified regions of the brain involved with recognizing emotions, right? No! The investigators could prove beyond any doubt that both findings were false positives resulting from noise in the instruments and had nothing to do with changes in blood flow. How could they be sure? Their experimental subject was a dead salmon! When they properly corrected for multiple comparisons to account for the thousands of brain regions at which they looked, neither finding was statistically significant.

MULTIPLE COMPARISONS IN MANY CONTEXTS

Multiple subgroups

Analyzing data divided into multiple subgroups is a common form of multiple comparisons. A simulated study by Lee and colleagues (1980) points out potential problems. These investigators pretended to compare survival following two "treatments" for coronary artery disease. Specifically, they studied a group of real patients with coronary artery disease and randomly divided them into two groups, A and B. In a real study, they would give the two groups different treatments and compare survival after treatment. In this simulated study, they treated all subjects identically but analyzed the data as if the two random groups were actually given two distinct treatments. As expected, the survival of the two groups was indistinguishable.

They then divided the patients in each group into six subgroups depending on whether they had disease in one, two, or three coronary arteries and whether the heart ventricle contracted normally. Because these variables are expected to affect survival of the patients, it made sense to evaluate the response to "treatment" separately in each of the six subgroups. Whereas they found no substantial difference in five of the subgroups, they did find a striking result among patients with three-vessel disease who also had impaired ventricular contraction. Of these patients, those assigned to treatment B had much better survival than those assigned to treatment A. The difference between the two survival curves was statistically significant, with a P value of less than 0.025.

If this were an actual study comparing two alternative treatments, it would be tempting to conclude that treatment B is superior for the sickest patients and recommend it to those patients in the future. But in this study, the random assignment to treatment A or treatment B did not alter how the patients were actually treated, so we can be absolutely, positively sure that the survival difference was a coincidence.

It is not surprising that the authors found one low P value out of six comparisons. Figure 17.1 shows there is a 26% chance that at least one of six independent comparisons will have a P value of less than 0.05 even if all null hypotheses are true.

For analyses of subgroups to be useful, it is essential that the study design specify all subgroups analyses, including methods to be used to account for multiple comparisons.

Multiple sample sizes

If you perform an experiment and the results aren't quite statistically significant, it is tempting to run the experiment a few more times (or add a few more subjects) and then analyze the data again with the larger sample size. If the results still aren't significant, then you might do the experiment a few more times (or add more subjects) and then reanalyze again. The problem is that you can't really interpret the results when you informally adjust sample size as you go.

If the null hypothesis of no difference is true, the chance of obtaining a statistically significant result using that informal sequential approach is far higher than 5%. In fact, if you carry on that approach long enough, every single experiment will eventually reach a statistically significant conclusion even if the null hypothesis is true. (Of course, "long enough" might be very long indeed, exceeding your budget or even your life span!)

The problem with this approach is that you continue collecting data only when the result is not statistically significant and stop when the result is statistically significant. If the experiment were continued after reaching significance, adding more data might then result in a conclusion, again, that the results are *not* statistically significant. But you'd never know this, because you would have stopped once significance was reached. If you keep running the experiment when you don't like the results but stop the experiment when you do, you can't interpret the results at face value.

Statisticians have developed rigorous ways to dynamically adjust sample size. Look up *sequential data analysis* or *adaptive sample size methods*. These methods use more stringent criteria to define significance in order to make up for the multiple comparisons. Without these special methods, you can't interpret the results unless the sample size is set in advance.

Multiple end points and misreporting which end point is "primary"

When conducting clinical studies, investigators often collect data on multiple clinical outcomes. When planning the study, investigators should specify the outcome they care most about, called the *primary outcome.* The other outcomes are called *secondary outcomes.* By now, you should realize why it is so important that the primary outcome be carefully defined in advance. If many outcomes are measured, you'll likely obtain statistical significance in at least one outcome just by chance. The analysis of secondary outcomes can strengthen the scientific argument of a paper and lead to new hypotheses to study. But statistical analyses must be focused on the primary outcome chosen in advance.

You can't draw proper inferences from clinical studies unless you can be sure that the investigators chose the primary outcome as part of the study plan—and stuck with that decision when writing up the results. Unfortunately, several studies have compared study protocols (plans) with the published results and discovered that investigators often misreport their choice of primary outcome (Chan et al., 2004; Vedula et al., 2009; Vera-Badillo et al., 2013).

Multiple definitions of groups

When comparing two groups, the groups must be defined as part of the study design. If the groups are defined by the data, many comparisons are being made implicitly, and the results cannot be interpreted.

Austin and Goldwasser (2008) demonstrated this problem when they looked at the incidence of hospitalization for heart failure in Ontario, Canada, in 12 groups of patients defined by their astrological sign (determined by birthday). People born under the sign of Pisces happened to have the highest incidence of heart failure. After seeing this association, the investigators did a simple statistics test to compare the incidence of heart failure among those born under Pisces with the incidence of heart failure among those born under all other 11 signs (combined into one group). Taken at face value, this comparison showed that the people born under Pisces have a statistically significant higher incidence of heart failure than do people born under the other 11 signs (the P value was 0.026).

The problem is that the investigators didn't really test a single hypothesis; they implicitly tested 12. They only focused on Pisces *after* looking at the incidence of heart failure for people born under all 12 astrological signs. So it wasn't fair to compare that one group against the others without considering the other 11 implicit comparisons. If you properly account for multiple comparisons, these data show no association between astrological sign and disease.

Multiple ways to preprocess the data

Scientists commonly do not analyze the direct experimental measurements but instead first process the data by transforming, normalizing, smoothing, adjusting (for other variables), or removing outliers (see Chapter 21). These kinds of manipulations can make lots of sense when done in a careful, preplanned way. However, you engage in multiple comparisons if you repeat the statistical analyses on data processed in multiple ways and then report analyses on the form that gives you the smallest P value.

Multiple analyses

This story told by Vickers (2006) is only a slight exaggeration:

> STATISTICIAN: Oh, so you have already calculated the P value?
> SURGEON: Yes, I used multinomial logistic regression.
> STATISTICIAN: Really? How did you come up with that?
> SURGEON: Well, I tried each analysis on the SPSS drop-down menus, and that was the one that gave the smallest P value. [SPSS is a commonly used statistics program.]

If you analyze your data many ways—perhaps first with a t test, then with a nonparametric Mann-Whitney test, and then with a two-way analysis of variance (adjusting for another variable)—you are performing multiple comparisons.

P-hacking

As already mentioned in Chapter 14, Simmons, Nelson, and Simonsohn (2011) pointed out the dangers of *P-hacking*, which are attempts by investigators to lower the P value by trying various analyses or analyzing subsets of data. When done without a plan and without a method to correct for multiple comparisons, all the forms of multiple comparisons discussed in this chapter are forms of P-hacking.

HOW TO CORRECT FOR MULTIPLE COMPARISONS

Familywise error rate

When each comparison is made individually and without any correction for multiple comparisons, the traditional 5% significance level applies to each individual comparison. This is therefore known as *the* per-comparison error rate, and it is the chance that random sampling will lead *this particular comparison* to an incorrect conclusion that the difference is statistically significant when *this particular null hypothesis* is true.

When you correct for multiple comparisons, the significance level is redefined to be the chance of obtaining *one or more* statistically significant conclusions if *all* of the null hypotheses in the family are actually true. The idea is to make a stricter threshold for defining significance. If α is set to the usual value of 5% and all of the null hypotheses are true, then the goal is to have a 95% chance of obtaining zero statistically significant results and a 5% chance of obtaining one or more statistically significant results. This 5% chance applies to the entire family of comparisons performed in the experiment, so it is called a *familywise error rate.* or the *per-experiment error rate.*

Bonferroni correction

The simplest approach to achieving a familywise error rate is to divide the value of α (often 5%) by the number of comparisons, then define a particular comparison as statistically significant only when its P value is less than that ratio. This is called a *Bonferroni correction.*

Imagine that an experiment makes 20 comparisons. If all 20 null hypotheses are true and there are no corrections for multiple comparisons, about 5% of these comparisons are expected to be statistically significant, and there is about a 64% chance of obtaining one or more statistically significant results (see Table 17.1).

With the Bonferroni correction, a result is only declared to be statistically significant when its P value is less than 0.05/20, or 0.0025. This ensures that if all the null hypotheses are true, there is about a 95% chance of seeing no statistically significant results among all 20 comparisons and only a 5% chance of seeing one or more statistically significant results (see Table 17.1). The 5% significance level applies to the entire family of comparisons rather than to each of the 20 individual comparisons.

Similarly, confidence intervals can be made wider (using the Bonferroni method) so that the 95% confidence level (or whatever you pick) applies to the entire family of confidence intervals, not just one. If this is done, you can be 95% confident that all the confidence intervals include the true population value, leaving a 5% chance that one or more confidence intervals do not include the true population value.

Note a potential point of confusion: The value of α (usually 0.05) applies to the entire family of comparisons, but a particular comparison is declared to be statistically significant only when its P value is less than α/K (where K is the number of comparisons).

Example of Bonferroni corrections: Genome-wide association studies

Genome-wide association studies use the Bonferroni correction to cope with a huge number of comparisons. These studies look for associations between diseases (or traits) and genetic variations (usually single nucleotide polymorphisms). They compare the prevalence of many (up to 1 million) genetic variations between many (often tens of thousands) cases and controls. To correct for multiple comparisons, the threshold is set using the Bonferroni correction. Divide α—in this case, 0.05—by the number of comparisons—say, 1 million—to calculate the threshold, which is 0.00000005 (5×10^{-8}).

NUMBER OF SIGNIFICANT COMPARISONS	NO CORRECTION	BONFERRONI
Zero	35.8%	95.1%
One	37.7%	4.8%
Two or more	26.4%	0.1%

Table 17.1. How many significant results will you find with 20 independent comparisons with α set to its conventional value of 0.05?

A tiny P value (less than 0.00000005) in a genome-wide association study testing one million genes is evidence that the prevalence of a particular genetic variation differs in the two populations you are studying.

The cost of correcting for multiple comparisons

Recall the definitions of Type I and II errors in Chapter 14. You've made a Type I error when your analysis leads you to conclude that a difference is statistically significant when in fact there is no such difference and your results are simply due to unlucky random sampling. You've made a Type II error when your analysis leads you to conclude that an observed effect is not statistical significance when there truly is an effect present that your data missed.

When you correct for multiple comparisons, you reduce the chance of making a Type I error. The preset chance of a Type I error (usually 5%) applies to the entire family of comparisons, not to each comparison individually (as it would if you did not correct for multiple comparisons). But the cost of this correction is an increase in the chance of making a Type II error. By making it harder to mistakenly find "statistically significant" results by chance, you've also made it harder to find real results.

When you make a family of confidence intervals so that the 95% confidence level applies to the entire family, the cost is that the intervals are wider than they would be if the 95% confidence level applied to each interval individually.

LINGO

What, exactly, is a *family of comparisons*? Usually, a family consists of all the comparisons in one experiment or all the comparisons in one major part of an experiment. But that definition leaves lots of room for ambiguity. When reading about results corrected for multiple comparisons, ask about how the investigators defined the family of comparisons.

COMMON MISTAKES

Mistake: Thinking that multiple comparisons are only an issue when you compare several groups

Multiple comparisons are not just a follow-up test after comparing several groups. The problem of multiple comparisons comes up when your sample size is determined dynamically, when multiple end points are studied, when groups can be defined in multiple ways, when there are multiple ways to preprocess the data before statistical analyses, when the investigator has tried several alternative analyses, when the investigator didn't choose the analysis until after viewing the data, and more.

Mistake: Not recognizing P-hacking

Statistical analyses can only be interpreted at face value when all the analysis steps were decided in advance. When results are not what investigators want or expect, many will increase the sample size, change the analysis method, choose subsets to analyze, and so on. This is known as P-hacking, and the results of P-hacking cannot be interpreted.

Q & A

Since most texts don't mention multiple comparisons, it seems that this is a problem that comes up rarely. Right?

No, multiple comparisons are everywhere. You can't interpret statistical results sensibly without always asking about the number of comparisons that were made.

Are corrections for multiple comparisons always needed?

No. Some statisticians recommend that investigators never correct for multiple comparisons but instead report uncorrected P values and confidence intervals with a clear explanation that no mathematical correction was made for multiple comparisons (Rothman, 1990). The people reading the results must then informally account for multiple comparisons. If all the null hypotheses are true, you can expect 5% of the comparisons to have uncorrected P values of less than 0.05. Compare this number with the actual number of small P values.

This approach requires that all analyses be planned and that all planned analyses be conducted and reported. These simple guidelines are often violated.

Another situation where corrections for multiple comparisons are not essential is clinical trials that clearly define, as part of the study protocol, that one outcome is primary. This is the key outcome on which the conclusion of the study is based. The study may include other, secondary comparisons, but those are clearly labeled as secondary. The secondary analyses are often not corrected for multiple comparisons.

CHAPTER SUMMARY

- The problem of multiple comparisons shows up in many situations, including multiple end points, multiple time points, multiple subgroups, multiple geographical areas, multiple methods of preprocessing data, and multiple statistical analyses.
- If you make 13 independent comparisons with all null hypotheses true, there is about a 50% chance that one or more P values will be less than 0.05.
- Coping with multiple comparisons is one of the biggest challenges in data analysis.
- You can only interpret statistical results when you know how many comparisons were made.
- All analyses should be planned, and all planned analyses should be conducted and reported. These simple guidelines are often violated.
- One way to deal with multiple comparisons is to analyze the data as usual but fully report the number of comparisons that were made and let the reader account for the number of comparisons.
- The most common approach to multiple comparisons is to use the Bonferroni method so that the significance level applies to an entire family of comparisons rather than to each individual comparison. The corrected confidence intervals also apply to the entire family of comparisons.

CHAPTER 18

Statistical Power and Sample Size

AD HOC SEQUENTIAL SAMPLE SIZE DETERMINATION LEADS TO MISLEADING RESULTS

This approach seems appealing: Collect and analyze some data. If the results are statistically significant, stop. But if the results are not statistically significant, collect some more data, and reanalyze. Keep collecting more data until you obtain a statistically significant result or you run out of money, time, or curiosity.

Figure 18.1 is a simulation that demonstrates the problem with *ad hoc sequential sample size determination*. The simulated data were sampled from populations with Gaussian distributions and identical means and standard deviations. Samples of n = 5 in each group were compared with an appropriate statistical test (the unpaired t test), and the resulting P value (0.406) is plotted at the left of the graph. Note that the Y axis is logarithmic to focus attention on the small P values. Then one more value was added to each group (n = 6), and the t test was run again with the larger sample sizes. That P value was 0.825. Then one more value was added to each group (n = 7), and the P value was computed again (0.808). This was continued up to n = 75.

Someone viewing these results would have been very misled if the investigator had stopped collecting data the first time the P value dropped below 0.05 and simply reported that the P value was 0.0286 and the sample size was 21 in each group. It is impossible to interpret results if an ad hoc method is used to choose sample size. In fact, if you could extend the sample size out to infinity, you would *always* eventually see a P value of less than 0.05 even if the null hypothesis were true. In practice, of course, you may run out of money, time, or patience before you obtain a P value less than 0.05. Even so, the chance of obtaining a P value less than 0.05 under the null hypothesis is much higher than 5%. Sample size must be determined in advance to obtain results that can readily be interpreted.

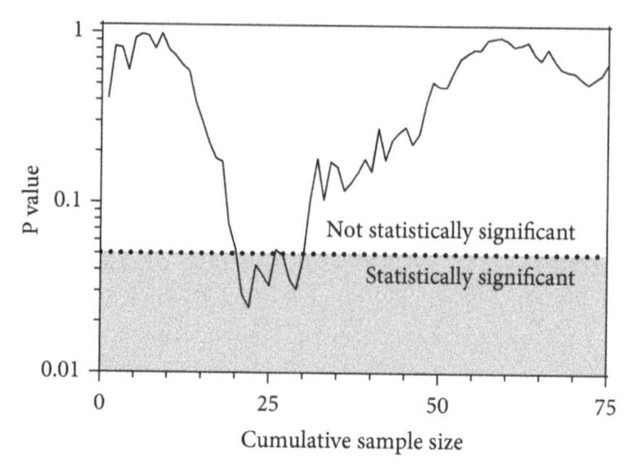

Figure 18.1. Beware of ad hoc sample size decisions.

This figure shows a simulation of a common mistake in data analysis. The (simulated) investigator started with n = 5 in each group and used a t test to determine a P value. The P value (plotted on the logarithmic Y axis) was (much) greater than 0.05, so the investigator increased the sample size in each group to six, and then to seven, eight, nine, and so on. All the values analyzed were sampled from the same Gaussian distribution, so the null hypothesis is known to be true. The figure shows what would have happened had the study continued to include n = 75 in each group. However, the investigator probably would have stopped at 21 in each group, when the result happened to hit statistical significance. Unless the analyses are done with special methods, P values simply cannot be interpreted when the sample size isn't determined in advance. Note that the values in this figure were calculated via simulation with random numbers. Another simulation would have looked different, but the main point would persist: The cumulative P value bounces around a lot, especially with small sample sizes.

THE FOUR QUESTIONS

To calculate necessary sample size, you must first answer four questions:

- **What is the size of the effect you are looking for?** It takes a larger sample size to find a small effect than it takes to find a large effect.
- **What is the significance level?** If you want to use a smaller (stricter) significance level, you'll need a larger sample size.
- **How much variability?** If you expect lots of variation among the data, you'll need a larger sample size than if you expect little variation.
- **How much power?** If there really is an effect (difference, relative risk, correlation, etc., which we will call *effect*) of a specified magnitude in the overall population, how sure do you want to be that you will find a statistically significant result in your proposed study? The answer is the *power* of the experiment. If you want a lot of power, you need a large sample size.

INTERPRETING A SAMPLE SIZE STATEMENT

Clinical trials usually compute sample size rigorously, and include a statement that reads something like this:

> We chose to study 313 subjects in each group in order to have 80% power to detect a 33% reduction in the recurrence rate from a baseline rate of 30% with a significance level of 0.05 (two tailed).

To understand this statement, you have to understand the answer to all four questions that were used to compute the sample size.

- Effect size. Here, the investigators were looking for a 33% relative reduction in recurrence rate (from 30% down to 20%).
- Significance level. This study defined α to its traditional value of 0.05, so P values of less than 0.05 are defined to be statistically significant.
- Variability. When comparing means, the required sample size depends on the expected value of the standard deviation. If there is a lot of variation (high standard deviation), you'll need larger samples. When comparing proportions (as in the example here), the required sample size depends on how far the proportions are expected to be from 50%. If you expect the outcome to occur about half the time, the sample size must be larger than if you expect the proportion to be closer to 0% or 100%. Here, the investigators estimated that the baseline recurrence rate (which they are hoping to reduce) is 30%.
- Power. The example study used a power of 80%. That means that if the effect the investigators wanted to see really exists, there was an 80% chance that the study would have resulted in a statistically significant result, leaving a 20% chance that the study would have determined that the difference was not statistically significant. If the investigators had wanted more statistical power, they would have needed a larger sample size.

Let's restate the sample size statement. The investigators assumed that the recurrence rate is normally 30% and asked whether the new treatment reduced this rate by one-third or more, down to 20% or less. They chose the usual significance level of 5% and a power of 80%. This means that if the true effect is a one-third reduction in recurrence rate, they will have an 80% chance of obtaining a statistically significant result (defined as $P < 0.05$). Given these decisions, a computer program (or simple equation) calculates that they need 313 subjects in each group.

A CALCULATION OR A NEGOTIATION?

The quotation above makes it seem that the investigator answered the four questions, and then used these to compute the sample size. In fact, the process may actually have been more like a negotiation. The investigators may have first said they were looking for a 10% change in recurrence rates and were horrified at the enormous number of subjects that would have been required to produce such a result. They then may have changed their goal until the needed sample size seemed "reasonable." They may also have fussed with the significance level and desired power. Such an approach is not cheating. It is a smart way to decide on sample size, so long as the sample size is chosen before you collect data. Parker and Berman (2003) point out that the goal often isn't to calculate the number of subjects needed but rather to answer the question "If I use n subjects, what information can I learn?"

In some cases, the calculations may convince you that it simply is not possible to learn what you want to know with the number of subjects you are able to use. This realization can be very helpful. It is far better to cancel such a study in the planning stage than to waste time and money on a futile experiment that won't have sufficient power. If the experiment involves any clinical risk or expenditure of public money, performing an study with too few subjects can even be considered unethical.

AN ANALOGY TO UNDERSTAND STATISTICAL POWER

If you find the concept of statistical power hard to understand, here is a silly analogy that might help (Hartung, 2005): You send your child into the basement to find a tool. He comes back and says, "It isn't there." What do you conclude? Is the tool there or not?

There is no way to be sure, so the answer must be a probability. If the tool really is in the basement, what is the chance your child would have found it? In other words, what was the power of your child's search? The answer depends on three factors (summarized in Table 18.1).

- How long did he spend looking? If he looked for a long time, he is more likely to have found the tool than if he looked for a short time. The time spent looking for the tool is analogous to the sample size. An experiment with a large sample size has high power to find an effect, while an experiment with a small sample size has less power.
- How big is the tool? It is easier to find a snow shovel than a tiny screwdriver used to fix eyeglasses. The size of the tool is analogous to the size of the effect you are looking for. An experiment has more power to find a big effect than a small one.

	TOOLS IN BASEMENT	STATISTICS
High power	• Spent a long time looking • Large tool • Organized basement	• Large sample size • Looking for large effect • Little scatter
Low power	• Spent a short time looking • Small tool • Messy basement	• Small sample size • Looking for small effect • Lots of scatter

Table 18.1. The analogy between searching for a tool and statistical power.

- How messy is the basement? If the basement is a real mess, he is less likely to find the tool than if the basement is carefully organized. The messiness is analogous to the experimental scatter. An experiment has more power when the data have little variation and less power when the data are very scattered.

SAMPLE SIZE AND THE MARGIN OF ERROR OF THE CONFIDENCE INTERVAL

Sample size calculations are usually presented in terms of statistical hypothesis testing. But they can also be recast in terms of confidence intervals. The idea is simple. Everything else being equal, a larger sample size produces a narrower confidence interval. If you can state the maximum acceptable width of the confidence interval, then you can calculate how many subjects you need. You can use the same calculations, programs, and tables as those used for computing the sample size needed to achieve statistical significance. Instead of entering the smallest difference you wish to detect, enter the largest acceptable margin of error, which equals half the width of the entire confidence interval.

LINGO

The term *power* sometimes gets stretched a bit beyond its technical definition explained above. Some investigators refer to a sample size calculation as a *power analysis* and then write about how the study is *powered to detect* a certain effect size. And when criticizing a published study, scientists may refer to it as being *underpowered*. That just means it has too little power to detect some effect worth detecting, which is a fancy way to say the sample size was too small.

Sample size calculations are often expressed as determining the number of subjects you need. This terminology makes sense when your experimental unit is a person or an animal. But the same calculations can be used no matter what your experimental unit. If you are doing work with cultured cells, your experimental unit might be the number of plates of cells you need or the number of wells (in a multiwell plate) you need.

COMMON MISTAKES

Mistake: Computing sample size based on published effect sizes without thinking

All sample size calculations require you to specify how large of an effect you are looking for. It takes a huge sample size to reliably detect a tiny effect but a smaller sample size to detect a huge effect. However, specifying the effect size is not easy. Many people prefer to use results from a prior study (or a pilot study) and then calculate a sample size large enough to detect an effect that large. The problem with this approach is that you might care about an effect that is smaller than the published (or pilot) effect size. You should compute the sample size required to detect the smallest effect worth detecting, not an effect size that someone else has published (or that you determined in a pilot study).

Mistake: Computing sample size based on standard effect sizes without thinking

In order to compute sample size, you need to specify an effect size. To avoid the difficulty of defining the minimum effect size they want to detect, many are tempted to pick a standard effect size instead. They use standard values for the significance level and power, so why not use a standard effect size too?

In a book that is almost considered the bible of sample size calculations in the social sciences, Cohen (1988) defines some *standard effect sizes*. He limits these recommendations to the behavioral sciences (his area of expertise), however, and warns that all general recommendations are more useful in some circumstances than others. Here are his guidelines for an unpaired t test:

- A "small" difference between means is equal to one-fifth the standard deviation.
- A "medium" effect size is equal to half the standard deviation.
- A "large" effect is equal to 0.8 times the standard deviation.

If you are having trouble deciding the effect size for which you are looking (and therefore are stuck and can't determine a sample size), you can decide whether you are looking for a small, medium, or large effect and then use Cohen's definitions.

Lenth (2001), however, argues that you should avoid these "canned" effect sizes, and I agree. You must decide how large a difference you care to detect based on understanding the experimental system you are using and the scientific questions you are asking. Cohen's recommendations seem a way to avoid thinking about the goals of the experiment. If you choose standard definitions of α (0.05), power (80%), and effect size (see above), then there is no need for any calculations. If you are comparing two groups (with a t test), you will need a sample size of 26 in each group to detect a large effect, 65 in each group to detect a medium effect, and 400 in each group to detect a small effect. Choosing a standard effect size is really the same as picking a standard sample size.

Mistake: Failing to plan for dropouts

Sample size calculations tell you the number of values you need in the analyses. Sometimes, however, subjects drop out of the study (or need to be excluded from the analysis for some reason), so you'll need to start with more than the computed number. For example, if you expect about 20% of the subjects to drop out, you should start with 20% more subjects than calculated.

Mistake: Using standard values for significance level and power without thinking

Sample size calculations require you to specify the significance level you wish to use as well as the power. Many simply choose the standard values: a significance level (α) of 5% ($P < 0.05$ is significant) and a power of 80%. It is better to choose values based on the relative consequences of Type I and Type II errors.

Q & A

Does 80% power mean that 80% of the subjects would be improved by the treatment and 20% would not?

No. Power is the probability that a proposed study will reach a conclusion assuming a true population difference (or effect). Power has nothing to do with the fraction of patients who benefit from a treatment.

How much power do I need?

Sample size calculations require you to choose how much power the study will have. If you want more power, you'll need a larger sample size. Often, power is set to a standard value of 80%. Ideally, the value should be chosen to match the experimental setting, the goals of the experiment, and the consequences of making a Type II error (see Chapter 15).

My program asked me to enter β. What's that?

Some sample size programs ask you to enter β instead of power. β equals 1.0 minus the power. A power of 80% means that if the effect size is what you predicted, your experiment has an 80% chance of producing a statistically significant result and thus a 20% chance of obtaining a not significant result. In other words, there is a 20% chance that you'll make a Type II error (missing a true effect of a specified size). β therefore equals 20%, or 0.20.

Why do I have to specify the effect size for which I am looking? What if I want to detect any size effect?

All studies will have a small power to detect tiny effects and a large power to detect enormous effects. You can't calculate power without specifying the effect size for which you are looking.

By how much do you need to increase the sample size to detect an effect size half as big?

A general rule of thumb is that increasing the sample size by a factor of 4 will cut the effect size that can be detected (and the expected width of the confidence interval) by a factor of 2. (Note that 2 is the square root of 4.)

What if I want to detect an effect size one-quarter as wide?

Increasing the sample size by a factor of 16 will allow you to detect an effect size one quarter as large. (Note that 4 is the square root of 16.)

CHAPTER SUMMARY

- Ad hoc sequential sample size determination is commonly done but not recommended. You can't really interpret P values or confidence intervals when the sample size was determined that way.
- The power of an experimental design is the probability that you will obtain a statistically significant result assuming a certain effect size in the population.
- The power of an experiment depends on the sample size, the variability, the significance level, and the hypothetical effect size.
- Power should be computed based on the minimum effect size that would be scientifically or clinically worth detecting. Once the study is complete, it is not helpful to compute the power that study had to detect the effect size that was actually observed.

- Sample size calculations require that you specify the significance level, the effect size for which you are looking, the scatter (standard deviation) of the data or the expected proportion, and the desired power.
- You need larger samples when you are looking for a small effect, when the standard deviation is large, and when you desire lots of statistical power.
- It is often more useful to ask what you can learn from a proposed sample size than to compute a sample size from the four variables listed in the previous bullet point.

CHAPTER 19

Commonly Used Statistical Tests

ASSUMPTIONS SHARED BY ALL STANDARD STATISTICAL TESTS

All statistical tests rely on a standard set of assumptions already discussed in Chapter 10 (and elsewhere).

Assumption: Random (or representative) sample

Statistical tests are based on the assumption that your samples were randomly selected from the populations. In many cases, this assumption is not strictly true. You can still interpret the confidence interval as long as you assume that your samples are representative of the populations from which they are drawn.

Assumption: Independent observations

Statistical results are only valid when all subjects are sampled from the same population and each has been selected independently of the others. Selecting one member of the population should not change the chance of selecting any other member.

Assumption: Accurate data

Statistical results are only valid when each value is measured correctly. If there are systematic errors in data collection, the results of statistical comparisons won't be useful.

Assumption: Assessing the event you really care about

Statistical tests allow you to generalize from the sample to the population for the event or variable that you tabulated. Sometimes, though, you really care about a different variable or event. Extrapolating from the measured variable (e.g., surrogate marker) to the variable you really care about (e.g., disease impact) requires scientific judgment and, occasionally, a leap of faith. Statistics can't help.

COMPARING A CONTINUOUS VARIABLE MEASURED IN TWO GROUPS

The choice of a test to compare two continuous data sets depends on whether you wish to use a parametric or a nonparametric approach and whether the experimental design includes matching or pairing.

Parametric versus nonparametric tests

Many statistical tests assume that continuous data are sampled from a Gaussian distribution (see Chapter 8). Tests that are based on an assumption about the distribution of values in the population (usually that the population is Gaussian) are called *parametric tests*.

Nonparametric tests make few assumptions about the distributions of the populations and so do not assume sampling from a Gaussian distribution. The most popular forms of nonparametric tests are based on a really simple idea: Rank the values from low to high, and then analyze only those ranks, ignoring the actual values. These kinds of tests are used for continuous data without assuming any particular distribution, and they are robust in the presence of outliers (see Chapter 21). These kinds of tests are also used with data that are already ranked, such as the scale used to rank pain from 1 to 10 or reviews that rank movies from 1 to 5 stars. (Note that the adjective *nonparametric* is used to describe a statistical test. It is never used to describe a variable or a data set.)

Paired versus unpaired tests

Special tests are needed when the experimental design involves pairing or matching, including the following situations:

- A variable is measured in each subject before and after an intervention.
- Subjects are recruited as pairs and matched for variables such as age, postal code, or diagnosis. One of each pair receives one intervention, whereas the other receives an alternative treatment.
- Twins or siblings are recruited as pairs. Each receives a different treatment.
- Each run of a laboratory experiment has a control and a treated preparation handled in parallel.
- A part of the body on one side is treated with a control treatment, and the corresponding part of the body on the other side is treated with the experimental treatment (e.g., right and left eyes).

Choosing a statistical test

Choosing a test depends on whether the experimental design involves pairing or matching and whether you wish to use a parametric or a nonparametric test:

- Unpaired, parametric. Use the *unpaired t test*, also called the *two-sample t test* (Table 19.1).
- Paired, parametric. Use the *paired t test* (Table 19.2).
- Unpaired, nonparametric. Use the *Mann-Whitney test* (Table 19.3).
- Paired, nonparametric. Use the *Wilcoxon matched pairs test* (Table 19.4).

Test	Unpaired t test
Also known as	Two-sample t test; independent samples t test
Goal	Compare two means
Example	Comparing pulse rate in people taking two different drugs
Additional assumption	Both data sets are sampled from Gaussian distributions with the same population standard deviation
Effect size	Difference between the two means
Confidence interval	Confidence interval of the difference between the two means
Null hypothesis	The two population means are identical
Question the P value answers	If the two population means are identical, what is the chance of observing such a large difference between means by chance alone in an experiment of this size?

Table 19.1. The unpaired t test.

Test	Paired t test
Also known as	Dependent t-test for paired samples
Goal	Compare a continuous variable before and after an intervention (or in matched samples)
Example	Comparing pulse rate before and after taking a drug
Additional assumption	The population of paired differences is Gaussian
Effect size	Mean of the paired differences
Confidence interval	Confidence interval of the mean of the paired differences
Null hypothesis	The population mean of the paired differences is zero
Question the P value answers	If there is no difference in the population, what is the chance of observing such a large difference between means by chance alone in an experiment of this size?

Table 19.2. The paired t test.

Test	Mann-Whitney test
Also known as	Wilcoxon rank sum test
Goal	Compare the average ranks or medians of two unpaired groups
Example	Comparing the self-reported pain score (on a scale of 0 to 10) of patients taking two different drugs
Additional assumption	For the interpretation regarding medians to be applicable, that both distributions have the same shape even if not Gaussian and the medians differ
Effect size	Difference between the two medians
Confidence interval	Confidence interval of the difference between the two medians
Null hypothesis	The two population distributions are identical
Question the P value answers	If the two populations are identical, what is the chance of observing such a large difference between mean ranks (or medians) by chance alone in an experiment of this size?

Table 19.3. The Mann-Whitney test.

Test	Wilcoxon matched pairs test
Also known as	Wilcoxon signed rank test
Goal	Compare the average ranks or medians of two paired groups
Example	Comparing the amount of skin inflammation (assessed on a scale of 0 to 10) between the right arm treated with one cream and the left arm treated with a different cream
Additional assumption	For the interpretation regarding the median to be applicable, that the distribution of differences is symmetrical even if not Gaussian
Effect size	Median of the paired differences
Confidence interval	Confidence interval of the median of the paired differences
Null hypothesis	The population median of the paired differences is zero
Question the P value answers	If there is no difference in the population, what is the chance of observing such a large median difference by chance alone in an experiment of this size?

Table 19.4. The Wilcoxon matched pairs test.

COMPARING A CONTINUOUS VARIABLE MEASURED IN THREE OR MORE GROUPS

A statistical test called *one-way analysis of variance* (abbreviated ANOVA) is used to compare the means of three or more groups (Table 19.5). The word *variance* refers to the method the test uses, not the hypotheses that it tests. One-way ANOVA compares means, not variances. It works by comparing the total variation of all the data (without paying attention to which group each value is in) to the variation within the groups. If the variation within the groups is much smaller than the total variation, then the group means must differ substantially. In this

Test	One-way analysis of variance (ANOVA)
Also known as	One-factor ANOVA
Goal	Compare three or more means
Example	Comparing the pulse rate of four groups of people, each group taking a different drug
Additional assumption	All data sets sampled from Gaussian distributions with the same population standard deviation.
Effect size	Fraction of the total variation explained by variation among group means (R^2; also called eta^2)
Null hypothesis	All population means are identical
Question the P value answers	If all population means are identical, what is the chance of observing such a large variation among sample means by chance alone in an experiment of this size?
Follow-up tests	See Tables 19.6 and 19.7

Table 19.5. One-way ANOVA.

Test	Tukey multiple comparison test
Also known as	Tukey-Kramer test
Goal	From three or more groups, compare every pair of means
Additional assumption	All data sets sampled from Gaussian distributions with the same population standard deviation
Effect size	A set of differences between every pair of means
Confidence interval	A set of confidence intervals for those differences (the confidence level, usually 95%, applies to the entire family of comparisons, not just the one)
Null hypothesis	All two population means are identical
Question the P value answers	If all population means are identical, what is the chance of observing at least two means with such a large difference between them by chance alone in an experiment of this size?
Notes	This test is usually performed after one-way ANOVA, but the interpretation of the results does not depend on the overall ANOVA results

Table 19.6. Tukey multiple comparison test.

case, the P value will be small, and you will reject the null hypothesis that all groups are sampled from populations with the same mean. Perhaps all the population means are distinct. Perhaps all but one of the population means are identical. Or perhaps some populations have one mean and some have another.

Following one-way ANOVA, it is common to run a multiple comparisons test. There are many such tests, but the most common methods are the *Tukey test* (Table 19.6), which is used to compare each group mean with every other group mean, and the *Dunnett test* (Table 19.7), which is used to compare each group mean with a control mean. The most important result from these tests are confidence intervals for the differences between pairs of means. These confidence intervals adjust for multiple comparisons, so the confidence level (usually 95%) applies to the entire family of comparisons, not just one comparison. They also can report which pairs of means are different enough to be considered statistically significant and which are not. Again, this accounts for the number of comparisons, so the significance level applies to the entire family of comparisons, not to each comparison individually.

COMPARING A BINARY VARIABLE ASSESSED IN TWO GROUPS

Examples of binary data include cure (yes/no) of acute myeloid leukemia within a specific time period, success (yes/no) of preventing motion sickness, and sex (male/female) of a baby. For each subject or experiment, the result has two possible outcomes. For a group of subjects or experiments, the result is expressed as a proportion. Compare two proportions with the *Fisher exact test* (Table 19.8). With large samples, the *chi-square test* can be used instead.

Test	Dunnett multiple comparison test
Goal	From three or more groups, compare each mean with the control mean
Additional assumption	All data sets sampled from Gaussian distributions with the same population standard deviation
Effect size	A set of differences between every mean and the control mean
Confidence interval	Confidence interval of the difference between a group mean and the control mean (the confidence level, usually 95%, applies to the entire family of confidence intervals, not just one)
Null hypothesis	All population means are identical
Question the P value answers	If the all population means are identical, what is the chance of observing at least one group mean this far or farther from the control mean in an experiment of this size?
Notes	This test is usually performed after one-way ANOVA, but the interpretation of the results does not depend on the overall ANOVA results. Because there are fewer comparisons than with the Tukey test, there is more power to detect differences.

Table 19.7. Dunnett multiple comparison test.

Test	Fisher exact test
Also known as	Fisher exact test of independence
Goal	Compare two proportions
Example	Tabulate the fraction of patients who get infections after surgery, and compare those for whom the surgeon and operating room personnel followed a checklist with those for whom the surgeon and operating room personnel did not
Effect size	Difference between two proportions (sometimes called attributable risk) or the ratio of two proportions (sometimes called relative risk)
Confidence interval	Confidence interval of the difference or ratio of the two proportions
Null hypothesis	The two proportions are identical in the population
Question the P value answers	If there is no difference between proportions in the population, what is the chance of observing such a large discrepancy by chance alone in an experiment of this size?
Notes	With large sample sizes, this test is nearly identical to the chi-square test, which is also called the Wald z test.

Table 19.8. Fisher exact test.

COMPARING SURVIVAL CURVES

"Survival" analyses are used to analyze data whenever you've measured time until an event occurs, and that event does not have to be death. Examples of survival data include months until death for prostate cancer patients, number of days

Test	Log-rank test
Also called	Mantel-Haenszel test
Goal	Compare two survival curves
Example	Compare survival of cancer patients treated with two therapies.
Additional assumptions	Many (reviewed in Chapters 5 and 29 of Motulsky, 2014)
Effect size	Ratio of median survival times or the hazard ratio
Confidence interval	Confidence interval of the ratio of median survivals times or confidence interval of the hazard ratio
Null hypothesis	The two populations have identical survival curves overall
Question the P value answers	If there is no difference in survival over time in the population, what is the chance of observing such a large discrepancy by chance alone in an experiment of this size?
Notes	The method to turn the raw data into a table of percent survival versus time was developed by Kaplan and Meier, and those plots are called Kaplan-Meier graphs.

Table 19.9. The log-rank test to compare survival curves.

until cold symptoms go away, and minutes until REM sleep begins. Raw survival data are first converted to graphs of percent survival versus time called *Kaplan-Meier graphs*, and two survival curves are computed using the *log-rank test* (Table 19.9).

CORRELATION AND REGRESSION

Correlation and regression are explained in Chapters 22 and 23.

LINGO

The term *nonparametric* characterizes an analysis method. A statistical test can be nonparametric or not. However, it makes no sense to describe data as being *nonparametric*, and the phrase "nonparametric data" should never ever be used.

Methods that analyze ranks are uniformly called nonparametric. Several of these tests are mentioned in this chapter. Beyond that, the definition gets slippery. Modern statistical methods including randomization, resampling and bootstrapping do not assume sampling from a Gaussian distribution, but they analyze the actual data, not the ranks. Some texts refer to these tests as nonparametric but others don't.

The chi-square, Fisher's exact test, and survival anlayses analyze discrete (counted) data, so don't assume a Gaussian distribution. Some texts call these tests nonparametric but others don't.

CHAPTER SUMMARY

- All statistical tests share a basic set of assumptions.
- Different statistical tests have been designed for different kinds of data.

CHAPTER 20

Normality Tests

TESTING FOR NORMALITY

Many statistical tests assume that the data were sampled from a population with a Gaussian distribution. Normality tests examine this assumption. All normality tests quantify the deviation of a distribution from the normal ideal and then compute a P value that answers the following question:

> If you randomly sample this many values from a Gaussian population, what is the probability of obtaining a sample that deviates from a Gaussian distribution as much as or more than this sample does?

If the P value from the normality test is smaller than the significance level you set (usually 5%), you reject that null hypothesis and accept the alternative hypothesis that the data are not sampled from a Gaussian population.

If the P value from a normality test is larger than the significance level, you conclude that the data are consistent with a Gaussian distribution. Of course, this doesn't prove the data were sampled from a Gaussian population. It only tells you that the deviation from the Gaussian ideal is not more than you'd expect to see from chance alone.

THE PROBLEMS WITH NORMALITY TESTS

The Gaussian distribution is an unreachable ideal

In almost all cases, we can be 100% sure that the data were not sampled from a population with an ideal Gaussian distribution. Why? Because an ideal Gaussian distribution extends infinitely in both directions and so includes both very large values and extremely negative values. In most scientific situations, however, there are constraints on the possible values. Pressures, concentrations, weights, and many other variables cannot have negative values, so they can never be sampled from populations with perfect Gaussian distributions. Other variables can be negative but have physical or physiological limits that don't allow extremely large values (or extremely low negative values). These variables also cannot follow a perfect Gaussian distribution.

Normality tests ask the wrong question

Because almost no variables you measure follow an ideal Gaussian distribution, why use tests that rely on the Gaussian assumption? Plenty of studies with simulated data have shown that the statistical tests based on the Gaussian distribution are useful when data are sampled from a population with a distribution that only approximates a Gaussian distribution. These tests are fairly robust to violations of the Gaussian assumption, especially if the sample sizes are large and equal.

When analyzing data, the question that matters is not whether the data were sampled from an ideal Gaussian population but whether the distribution from which they were sampled is close enough to the Gaussian ideal that the results of the statistical tests are still useful. Normality tests do not answer this question.

ALTERNATIVES TO ASSUMING A GAUSSIAN DISTRIBUTION

If you don't wish to assume that your data were sampled from a population with a Gaussian distribution, you have several choices:

- Identify another distribution from which the data were sampled. In many cases, you can then transform your values to create a Gaussian distribution. Most commonly, if the data come from a population with a lognormal distribution (see Chapter 9), you can transform all values to their logarithms.
- Ignore small departures from the Gaussian ideal. Statistical tests tend to be quite robust to mild violations of the Gaussian assumption, especially with large sample sizes.
- Identify and remove outliers (see Chapter 21).
- Switch to a nonparametric test that doesn't assume a Gaussian distribution (see Chapter 19).

LINGO

Don't make the common mistake of asking whether a particular data set is Gaussian.

The term *Gaussian* refers to a population, not a data set. Normality tests ask whether the data are consistent with the assumption that your data were sampled from a population with a Gaussian distribution.

Also don't make the common mistake of asking whether your data are parametric or nonparametric. The terms *parametric* and *nonparametric* refer to statistical tests, not data distributions.

Ask about whether tests are parametric or nonparametric. Ask whether samples are consistent with sampling from a Gaussian population.

COMMON MISTAKES

Mistake: Using a normality test to automate the decision of when to use a nonparametric test

Nonparametric tests (explained in Chapter 19) do not assume a Gaussian distribution, so it seems logical to use the results of a normality test to decide whether to use a nonparametric test. But in fact, the decision of when to use nonparametric tests is not straightforward, and reasonable people can disagree about when to use them. The results of a normality test are not very helpful in making that decision. The problem is that a normality test is designed to detect evidence that the distribution deviates from a Gaussian distribution, but it does not quantify whether that deviation is large enough to invalidate the usual tests that assume a Gaussian distribution.

Mistake: Using a normality test with tiny samples

Normality tests try to determine the distribution (population) from which the data were sampled. That requires data. With tiny samples (three or four values), it really is impossible to make useful inferences about the distribution of the population.

Mistake: Using a normality test without looking at a graph showing the distribution of the values

Look at the data, not just the results of statistical tests.

Q & A

Will every normality test give the same result?

No. The different tests quantify deviations from normality in different ways, so they will return different results.

How many values are required to run a normality test?

It depends on which test you choose. Some work with as few as five values.

Does a normality test decide whether the data are far enough from Gaussian to require a nonparametric statistical test?

No. It is hard to define what "far enough" means, and the normality tests were not designed with this aim in mind.

How useful are normality tests?

Not very. With small samples, the normality tests don't have much power to detect non-Gaussian distributions. With large samples, they can detect minor deviations from the Gaussian ideal, but these differences wouldn't have much effect on the validity of a statistical test.

What are the names of the most commonly used normality tests?

Some commonly used normality tests are the Shapiro-Wilk test, the D'Agostino-Pearson test, the Anderson-Darling test, and the Kolmogorov-Smirnov test.

Is *normalizing* a method to make the distribution closer to a normal or Gaussian distribution?

Normalizing has many meanings, but normalizing data often does not change the shape of the distribution.

CHAPTER SUMMARY

- The Gaussian distribution is an ideal that is rarely achieved. Few, if any, scientific variables completely follow a Gaussian distribution.
- Normality tests are used to test for deviations from the Gaussian ideal.
- A small P value from a normality test only tells you that the deviations from the Gaussian ideal are more than you'd expect to see by chance.
- Normality tests tell you nothing about whether the deviations from the Gaussian ideal are large enough to affect conclusions from statistical tests that assume sampling from a Gaussian distribution.

CHAPTER 21

Outliers

HOW DO OUTLIERS ARISE?

An *outlier*—also called an *anomalous, spurious, rogue, wild,* or *contaminated* observation—is a value so far from the others that it appears to have come from a different population. Outliers can occur for several reasons:

- Invalid data entry. The outlier may simply be the result of transposed digits or a shifted decimal point.
- Biological diversity. If each value comes from a different person or animal, some values may be far from the others.
- Random chance. By chance, some values may end up far from the others.
- Experimental mistakes. Most experiments have many steps, and a mistake could cause a value to be very wrong.
- Skewed distribution. In a lognormal distribution, large values far from the rest will occur commonly (see Chapter 9).

THE NEED FOR OUTLIER TESTS

The presence of an outlier can spoil many analyses, either by creating the appearance of differences (or associations, or correlations) or by blocking the discovery of real differences (or associations, or correlations).

The presence of outliers would seem to be obvious. And if this were the case, we could deal with outliers informally. But formal methods are needed to identify outliers to prevent two problems:

- We see too many outliers. Even though all the data in Figure 21.1 were sampled from a Gaussian distribution, some points just seem to be too far from the rest, and it is tempting to eliminate them.
- Experimenters are biased. Even if you try to be fair and objective, your decision about which values to remove as outliers will probably be influenced by the results you want to see.

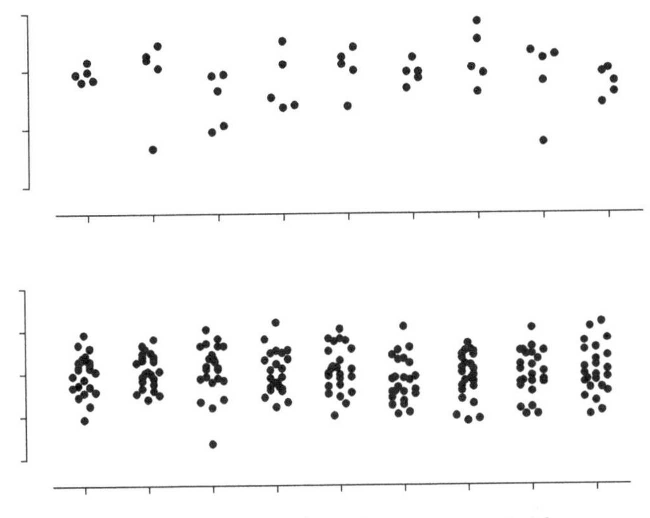

Figure 21.1. No outliers here. These data were sampled from Gaussian distributions.

All of these data sets were computer generated and sampled from a Gaussian distribution. The top graph shows nine samples with n = 5. The bottom graph shows nine samples with n = 21. Even though all the data were sampled from the same distribution, some points just seem too far from the rest to be part of the same distribution. They seem like real outliers—but they are not. The human brain is very good at seeing patterns and exceptions from patterns, but it is poor at recognizing random scatter.

FIVE QUESTIONS TO ASK BEFORE TESTING FOR OUTLIERS

Before using an outlier test, ask yourself these five questions about any suspiciously high or low values:

- Was there a mistake in data entry? If so, fix it.
- Is the extreme value really a code for missing values? Some programs require you to indicate missing values as something like 999. Make sure you entered these values and configured the program properly so that those entries don't get analyzed as if they were actual data.
- Did you notice a problem with the tube or well or animal generating that result during the experiment?
- Could the extreme values be a result of biological variability? If each value comes from a different person or animal, some correct values may be far from the others. This may be the most exciting finding in your data!
- Is it possible the distribution is not Gaussian? If so, you might be able to transform the values to make them Gaussian (you've seen a common

example in Chapter 9). Most outlier tests assume that the data (except the potential outliers) come from a Gaussian distribution.

THE QUESTION THAT AN OUTLIER TEST ANSWERS

If a value is much higher (or lower) than the others, and if you've answered no to all five questions above, two possibilities remain:

- The extreme value came from the same distribution as the other values and just happened to be larger (or smaller) than the rest. In this case, the value should not be treated specially.
- The extreme value was the result of a mistake. This could be something like bad pipetting, a voltage spike, or holes in filters. Or it could be a mistake in recording the value. These kinds of mistakes are not always noticed during data collection. Because including an erroneous value in your analyses will give invalid results, you should remove it. In other words, the value comes from a different population than the other values and is misleading.

The problem, of course, is that you can never be sure which of these possibilities is correct. Mistake or chance? No mathematical calculation can tell you for sure whether the extreme value came from the same or a different population than the others. An *outlier test*, however, can calculate a P value that answers the following question:

> If the values really are all sampled from a Gaussian distribution, what is the chance one value will be as far from the others as the extreme value that you observed?

If this P value is small, conclude that the outlier is not from the same distribution as the other values. Assuming you answered no to all five questions listed above, you have justification to exclude it from your analyses.

If the P value is high, you have no evidence that the extreme value came from a different distribution than the rest. But this does not prove that the value came from the same distribution as the others. All you can say is that there is no strong evidence that the value came from a different distribution.

IS IT LEGITIMATE TO REMOVE OUTLIERS?

It clearly is cheating to remove outliers in an ad hoc manner only those outliers get in the way of obtaining results you like. However it is not cheating when the decision to remove an outlier is based on criteria established before the data were collected and these criteria (along with the number of outliers removed) are reported when the data are published.

When an outlier test flags a value as an outlier, it is possible that a coincidence occurred—the kind of coincidence that happens in 5% (or whatever level you pick) of experiments even if the entire scatter is Gaussian. It is also possible that the value is a "bad" point. Which possibility is more likely? The answer depends on your experimental system:

- If your experimental system generates one or more bad points in a small percentage of experiments, eliminate the value as an outlier. It is more likely to be the result of an experimental mistake than to come from the same distribution as the other points.
- If your system is very pure and controlled, such that bad values almost never occur, keep the value. It is more likely that the value comes from the same distribution as the other values than that it represents an experimental mistake.

LINGO

The term *outlier* can be used inconsistently. Some authors use the term outlier to refer to an extreme value that appears to be far from the rest. Other times, the term *outlier* is only used to refer to values that have been flagged by a rigorous outlier analysis.

COMMON MISTAKES

Mistake: Not realizing that outliers are common in data sampled from a lognormal or some other skewed distribution

Most outlier tests are based on the assumption that all data—except the potential outlier(s)—are sampled from a Gaussian distribution. The results are misleading if the data were sampled from some other distribution.

Figure 21.2 shows values sampled from lognormal distributions. An outlier test, based on the assumption that most of the values were sampled from a Gaussian distribution, identified an outlier in three of the four data sets shown. But large values like these are not surprising in a lognormal distribution. Excluding these high values as outliers would be a mistake and lead to incorrect results. If you recognize that the data are lognormal, it is easy to analyze them properly. Transform all the values to their logarithms (right side of Figure 21.2), and those large values no longer look like outliers.

Mistake: Using a test designed to detect a single outlier when there are several outliers

Detecting multiple outliers is much harder than identifying a single outlier. The problem is that the presence of the second outlier can mask the first one so that neither is identified. Special tests are therefore needed. It is tempting to first run an outlier test, then remove the extreme value, and then run the outlier test again on the remaining values, but this method does a bad job of identifying multiple outliers.

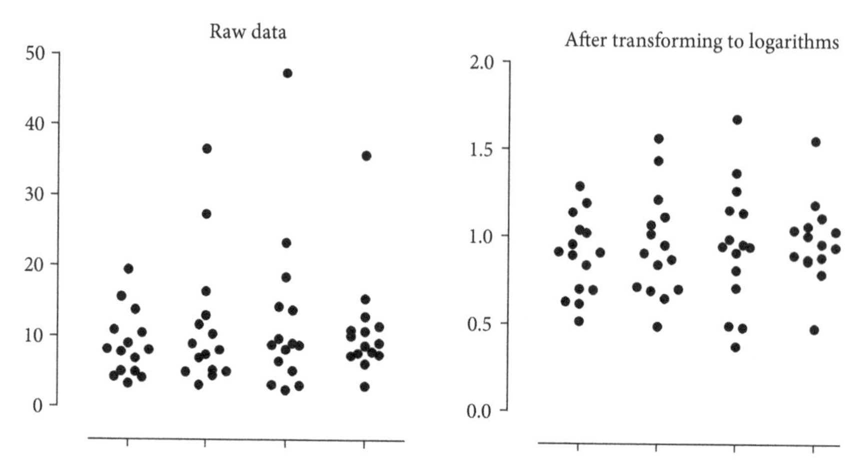

Figure 21.2. No outliers here. These data were sampled from lognormal distributions. (Left) The four data sets were randomly sampled by a computer from a lognormal distribution. A Grubbs outlier test found a significant ($P < 0.05$) outlier in three of the four data sets. (Right) The same values after being transformed to their logarithms. No outliers were found. Most outlier tests are simply inappropriate when values are not sampled from a Gaussian distribution.

Mistake: Eliminating outliers only when you don't get the results you want

If you are going to remove outliers from your data, it must be done in a systematic way, using criteria that have been set up in advance. It is not OK to first analyze the data and then only look to remove outliers when you don't get the results you want. That is a sure way to get invalid results.

Mistake: Truly eliminating outliers from your records

It can make sense to exclude outliers from graphs and analyses when you do so in a systematic way that you report along with your results. But the outliers should still be recorded in your notebook or computer files so that you have a full record of your work.

Q & A

What does it mean to remove or eliminate an outlier?

When an outlier is eliminated, the analyses are performed as if that value had never been collected. If the outliers are graphed, they are clearly marked. Outliers should not be erased from your lab notebook or computer files. Instead, the value of the outlier and the reason it was excluded should be recorded.

How should outlier removal be reported?

A scientific paper should state how many values were removed as outliers, the criteria used to identify the outliers, and whether those criteria were chosen as part of the experimental design. When possible, report the results computed in two ways, with the outliers included and with the outliers excluded.

Can it make sense to run statistical analyses two ways, with and without outliers?

Yes. This is commonly done. If the conclusions with the outliers removed are not very different from the results without the outliers removed, then no one needs to worry much about the outliers.

Can reasonable scientists disagree about how to deal with outliers?

Yes!

CHAPTER SUMMARY

- Outliers are values that lie very far from the other values in the data set.
- The presence of a true outlier can lead to misleading results.
- If you try to remove outliers manually, you are likely to be fooled. Random sampling from a Gaussian distribution creates values that often appear to be outliers but are not.
- Before using an outlier test, make sure the extreme value wasn't simply a mistake in data entry.
- Don't try to remove an outlier if it likely reflects true biological variability.
- Beware of multiple outliers. Most outlier tests are designed to detect one outlier and do a bad job of detecting multiple outliers. In fact, the presence of a second outlier can cause the outlier test to miss the first outlier.
- Beware of lognormal distributions. Very high values are expected in lognormal distributions, but these can be easily mistaken for outliers.

CHAPTER 22

Correlation

INTRODUCING THE CORRELATION COEFFICIENT

An example: Lipids and insulin sensitivity

Borkman and colleagues (1993) wanted to understand why insulin sensitivity varies so much among individuals. They hypothesized that the lipid composition of the cell membranes of skeletal muscle affects the sensitivity of the muscle to insulin.

In their experiment, they determined the insulin sensitivity of 13 healthy men by infusing insulin at a standard rate (adjusting for size differences) and quantifying how much glucose they needed to infuse to maintain a constant blood glucose level. Insulin causes the muscles to take up glucose, which in turn causes the level of glucose in the blood to fall, so people with high insulin sensitivity require a larger glucose infusion to maintain a constant level in the blood.

These investigators also took a small muscle biopsy specimen from each subject and measured its fatty acid composition. Here, we'll focus on the fraction of polyunsaturated fatty acids that have between 20 and 22 carbon atoms (%C20–22).

Table 22.1 shows the data, which are graphed in Figure 22.1. The graph shows a clear relationship between the two variables. Individuals whose muscles have more C20–22 polyunsaturated fatty acids tend to have greater sensitivity to insulin. The two variables vary together—statisticians say that there is a lot of *covariation*, or *correlation*. The results of a correlation analysis are shown in Table 22.2.

Correlation coefficient

The direction and magnitude of the linear correlation can be quantified with a *correlation coefficient* (abbreviated r). Its value can range from −1.0 to 1.0. If the correlation coefficient is 0.0, then the two variables do not vary together at all. If the correlation coefficient is positive, the two variables tend to increase or decrease together. If the correlation coefficient is negative, the two variables are inversely related—that is, as one variable tends to decrease, the other tends to increase. If the correlation coefficient is 1.0 or −1.0, the two variables vary together completely—that is, a graph of the data points forms a straight line.

Nonparametric correlation

Nonparametric tests (briefly introduced in Chapter 19) make few assumptions about the distribution of the populations. One nonparametric method for quantifying correlation is called the *Spearman rank correlation.* It separately ranks the X and Y values and then computes the correlation between the two sets of ranks. For the example data, the nonparametric correlation coefficient (abbreviated r_s) is 0.74 (the s in the subscript stands for Spearman). The P value, which tests the null hypothesis that there is no rank correlation in the overall population, is 0.0036. To distinguish the two ways to compute correlation, the usual method (which assumes Gaussian distributions) is said to calculate the *Pearson correlation coefficient,* while the nonparametric method computes the *Spearman correlation coefficient.*

Pearson correlation quantifies the linear relationship between X and Y. Even though correlation doesn't fit or plot a line, it essentially asks the degree to which the overall trend of the data follows a straight line. In other words, as X goes up, does Y go up a consistent amount?

In contrast, since Spearman correlation only sees the ranks of the values, it asks about the consistency of the monotonic relationship between X and Y. In other words, as X goes up, does Y go up as well (by any amount)?

ASSUMPTIONS: CORRELATION

You can calculate the correlation coefficient for any set of data, and it may be a useful descriptor of that data. However, interpreting the confidence interval and P value depends on several assumptions.

Assumption: Random sample

As with all statistical analyses, you must assume that subjects are randomly selected from, or at least representative of, a larger population.

Assumption: Independent observations

Correlation assumes that any random factor affects only one subject and not others. The relationship between all the subjects should be the same. In the %C20–22 fatty acids example, the assumption of independence would be violated if some of the subjects were related (e.g., siblings). It would also be violated if the investigators purposely chose some people with diabetes and some without or measured each subject on two occasions and treated the values as two separate data points.

Assumption: X values are not used to compute Y values

The correlation calculations are not meaningful if the values of X and Y are not measured separately. For example, it is not meaningful to calculate the correlation between a midterm exam score and the overall course score, because the midterm exam is one of the components of the overall course score.

Assumption: X values are not experimentally controlled

If you systematically control the X variable (e.g., concentration, dose, or time), you should calculate linear regression rather than correlation (see Chapter 23).

You will get the same value for r^2 and the P value. However, the confidence interval of r cannot be interpreted if the experimenter controlled the value of X.

Assumption: Both X and Y follow a Gaussian distribution

For Pearson (standard) correlation, the X and Y values must each be sampled from populations that follow a Gaussian distribution, at least approximately. If this assumption isn't true, the P value cannot be interpreted at face value. Nonparametric Spearman correlation does not make this assumption.

Assumption: All covariation is linear

The correlation coefficient is not meaningful if, for example, Y increases as X increases up to a certain point but then Y decreases as X increases further. Curved relationships are common but are not quantified with a Pearson correlation coefficient.

Assumption: No outliers

Calculation of the Pearson (but not the Spearman) correlation coefficient can be heavily influenced by one outlying point. Change or exclude that single point, and the results may be quite different. Outliers can influence all statistical calculations, but especially correlation. Look at graphs of the data before reaching any conclusion from correlation coefficients. Don't instantly dismiss outliers as bad points that mess up the analysis. The outliers may be the most interesting observations in the study!

LINGO

Correlation

When you encounter the word *correlation*, distinguish its strict statistical meaning from its more general usage. As used by statistics texts and programs, correlation quantifies the association between two continuous (interval or ratio) variables. However, the word is often used much more generally to describe the association of any two variables even if one or both of these variables are not continuous variables. It is not possible to compute a correlation coefficient to help you figure out whether survival times are correlated with choice of drug or whether antibody levels are correlated with sex.

Coefficient of determination

Coefficient of determination is just a fancy term for r^2. Most scientists and statisticians just call it *r square* or *r squared.*

COMMON MISTAKES

Mistake: Believing that correlation proves causation

Why do the two variables in our example correlate so well? There are four possible explanations:

- The lipid content of the membranes determines insulin sensitivity.
- The insulin sensitivity of the membranes somehow affects lipid content.
- Both insulin sensitivity and lipid content are under the control of some other factor, such as a hormone.
- The two variables don't correlate in the population at all, and the observed correlation in the sample is a coincidence.

You can never rule out the last possibility, but the P value tells you how unlikely the coincidence would be. In this example, you would observe a correlation this strong (or stronger) in 0.21% (the P value) of experiments if there was no correlation in the overall population.

You cannot decide which of the first three possibilities is correct by analyzing only these data. The only way to figure out which is true is to perform additional experiments in which you manipulate the variables. Remember that this study simply measured both values in a set of subjects. Nothing was experimentally manipulated.

Mistake: Focusing on the P value rather than the correlation coefficient

With large sample sizes, fairly small values of r can lead to a small P value and a conclusion that the correlation is statistically significant. This means you have evidence against the idea of zero correlation. But the nonzero correlation may be tiny and irrelevant. Always ask if the value of r (or r^2) is large enough to be considered interesting and worth pursuing. That is a scientific question, not a statistical one.

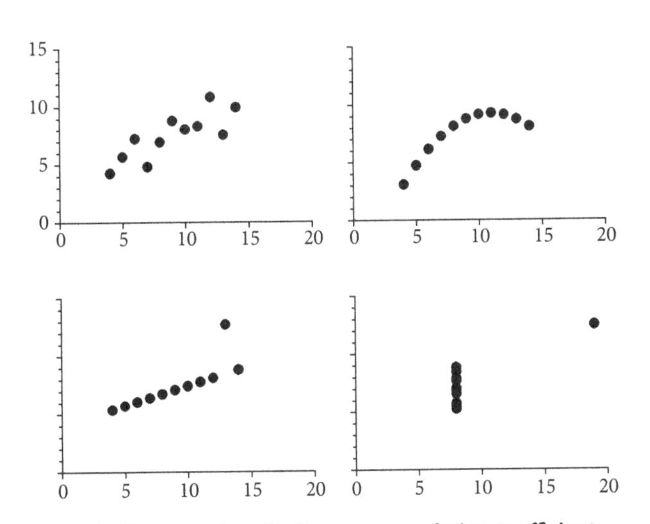

Figure 22.2. Four graphs with the same correlation coefficient.
In all four cases, the correlation coefficient (r) is 0.8164. Only in the first graph do the data appear to meet the assumptions of correlation.

Mistake: Interpreting the correlation coefficient without first looking at a graph

Don't try to interpret a correlation coefficient (or the corresponding P value) until you have looked at a graph of the data. Why? Look at Figure 22.2, which graphs four artificial data sets designed by Anscombe (1973). The correlation coefficients are identical (0.816), as are the P values (0.0022). But the data are very different.

Q & A

Do X and Y have to be measured in the same units to calculate a correlation coefficient?

X and Y do not have to be measured in the same units, but they can be.

In what units is r expressed?

None. It is unitless.

Can r be negative?

Yes. It is negative when one variable tends to go down as the other goes up. It is positive when one variable tends to go up as the other variable goes up.

Can correlation be calculated if all the X or all the Y values are the same?

No. If all the X or Y values are the same, it makes no sense to calculate correlation.

Why are there no best-fit lines in Figure 22.1?

Correlation quantifies the relationship but does not fit a line to the data. Chapter 23 explains linear regression, which finds the best-fit line, and reviews the differences between correlation and regression.

If all the X or Y values are transformed to new units, will r change?

No. Multiplying by a factor to change units (e.g., inches to centimeters, milligrams per milliliter to millimolar) will not change the correlation coefficient.

If all the X or Y values are transformed to their logarithms, will r change?

Yes. A logarithmic transformation (or any transformation that changes the relative values of the points) will change the value of r. However, the nonparametric Spearman correlation coefficient, which depends only on the rank order of the values, will not change.

If you interchange X and Y, will r change?

No. X and Y are completely symmetrical in calculating and interpreting the correlation coefficient.

If you double the number of points but r doesn't change, what happens to the confidence interval and the P value?

With more data points and the same value of r, the P value will be smaller, and the confidence interval of r will be narrower.

Is there a distinction between r^2 and R^2?

No. Upper- and lowercase mean the same thing in this context (correlation). However, the correlation coefficient, r, is always written in lowercase.

CHAPTER SUMMARY

- The correlation coefficient, r, quantifies the strength of the linear relationship between two continuous variables.
- The value of r is always between –1.0 and 1.0. It has no units.
- A negative value of r means that the trend is negative—that is, Y tends to go down as X goes up.
- A positive value of r means that the trend is positive—that is, X and Y go up together.

- The square of r equals the fraction of the variance that is shared by X and Y. This value, r^2, is sometimes called the *coefficient of determination.*
- Correlation is used when you measure both X and Y variables, but it is not appropriate if X is a variable that you manipulate.
- Correlation and linear regression are not the same. Correlation analysis does not fit or plot a line.
- You should always view a graph of the data when interpreting a correlation coefficient.
- Correlation does not imply causation. The fact that two variables are correlated does not mean that changes in one cause changes in the other.

CHAPTER 23

Simple Linear Regression

THE GOALS OF LINEAR REGRESSION

Recall the insulin example of Chapter 22. Curious to understand why insulin sensitivity varies so much among individuals, the investigators measured insulin sensitivity and the lipid content of muscle specimens obtained at biopsy in 13 men. You've already seen that the two variables (insulin sensitivity and the fraction of the fatty acids that are unsaturated with 20–22 carbon atoms [%C20–22]) correlate substantially.

The investigators concluded that differences in the lipid composition affect insulin sensitivity and proposed a simple model—namely, that insulin sensitivity is a linear function of %C20–22. In other words, as %C20–22 goes up, so does insulin sensitivity. This relationship can be expressed as an equation:

$$\text{Insulin sensitivity} = \text{Intercept} + \text{Slope} \cdot \%\text{C20–22}$$

Define the insulin sensitivity to be Y and the %C20–22 to be X, and the model is:

$$Y = \text{Intercept} + \text{Slope} \cdot X$$

We can't possibly know the true population value for the intercept or the slope. Given the sample of data, our goal is to find values for the intercept and slope that are most likely to be correct and quantify the precision with confidence intervals. More simply, we want to find the straight line that comes closest to the points on a graph. More precisely, we want to find the line that best predicts Y from X.

LINEAR REGRESSION RESULTS

Figure 23.1 and Table 23.1 shows the results of *linear regression* for our insulin sensitivity example.

The slope and its confidence interval

The best-fit value of the *slope* is 37.2 mg/m²/min. This means that when %C20–22 increases by 1.0, the average insulin sensitivity is expected to increase by

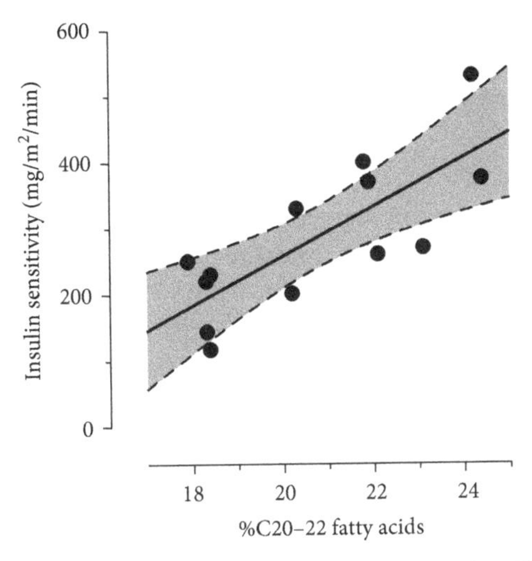

Figure 23.1. Linear regression of the %C20–22 versus insulin sensitivity data.

These are the same data shown in Figure 22.1. But this graph superimposes the best-fit line determined by linear regression as well as its 95% confidence band.

Slope	37.21
95% Confidence interval	16.75 to 57.67
Y intercept when X = 0.0	−486.5
95% Confidence interval	−912.9 to −60.17
R^2	0.5929
Sy.x (standard deviation of residuals)	75.90
P value	0.0021

Table 23.1. Linear regression results.

37.2 mg/m²/min. The 95% CI of the slope ranges from 16.8 to 57.7 mg/m²/min. Although the confidence interval is fairly wide, it does not include zero. In fact, it doesn't even come close to zero. This is convincing evidence of a strong relationship between lipid content of the muscles and insulin.

Some programs report the standard errors of the slope, which is 9.30 mg/m²/min. Confidence intervals are easier to interpret than standard errors, but the two are related. With large sample sizes, the confidence interval is approximated by the best-fit value plus or minus two standard errors.

The intercept

The best-fit value of the *intercept* is −486.5 mg/m²/min. This is the predicted value of the insulin sensitivity when the %C20–22 equals zero. In this example, the values of X range from about 18% to 24%. Extrapolating back to zero is not helpful or physiologically relevant. The intercept has a negative value, which is not biologically possible, because insulin sensitivity is assessed as the amount of glucose needed to maintain a constant blood level and therefore must be positive. These results tell us that the linear model cannot be correct when extrapolated way beyond the range of the data, even though it works fine in the range of our data.

Graphical results

Figure 23.1 shows the best-fit regression line defined by the values for the slope and the intercept as determined by linear regression.

The shaded area in Figure 23.1 shows the 95% *confidence bands* of the regression line, which combine the confidence intervals of the slope and the intercept. The best-fit line determined from this particular sample of subjects (solid black line) is unlikely to really be the best-fit line for the entire (infinite) population. If the assumptions of linear regression are true (which will be discussed later in this chapter), you can be 95% sure that the overall best-fit regression line lies somewhere within the shaded confidence bands.

The 95% confidence bands are curved but do not allow for the possibility of a curved (nonlinear) relationship between X and Y. The confidence bands are computed as part of linear regression, so they are based on the same assumptions as linear regression. The curvature simply is a way to enclose all possible straight lines, of which Figure 23.2 shows two.

The 95% confidence bands enclose a region that you can be 95% confident includes the true best-fit line (which you can't determine from a finite sample of data). But note that only six of the 13 data points in Figure 23.2 are included within the confidence bands. If the sample were much larger, the best-fit line would be determined more precisely, the confidence bands would be narrower,

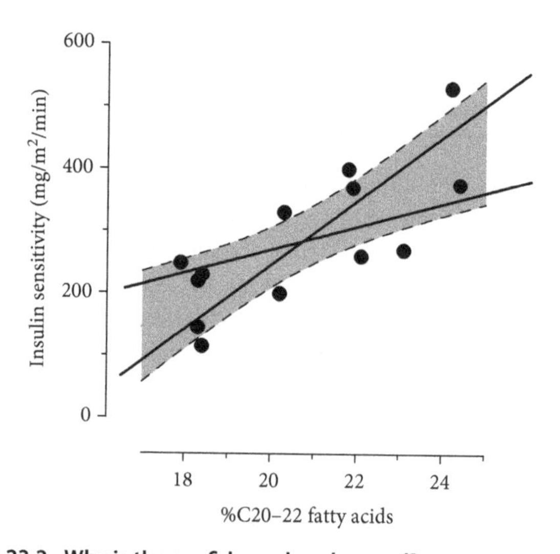

Figure 23.2. Why is the confidence band curved?

Given the assumptions of the analysis, you can be 95% sure that the (straight) true (population) line can be drawn within the shaded area. Two possible lines, within the confidence bands, are shown.

and a smaller fraction of data points would be included within the confidence bands. Note the similarity to the 95% CI for the mean, which does not include 95% of the values (see Chapter 10).

R^2

R^2 equals 0.59, which means that 59% of all the variance in insulin sensitivity can be accounted for by the linear regression model and the remaining 41% of the variance may be caused by other factors, measurement errors, biological variation, or a nonlinear relationship between insulin sensitivity and %C20–22. The value of R^2 is always between 0.0 (no linear relationship between X and Y, so the best-fit line is horizontal) and 1.0 (the graph of X vs. Y forms a perfect line).

P value

To interpret any P value, the null hypothesis must be stated. With linear regression, the null hypothesis is that no linear relationship between insulin sensitivity and %C20–22 exists, so the best-fit line in the overall population will be horizontal with a slope of zero. In this example, the 95% CI for slope does not include zero (and does not even come close), so the P value must be less than 0.05. In fact, it is 0.0021. The P value answers the following question:

> If that null hypothesis is true, what is the chance that linear regression of data from a random sample of subjects will have a slope as far or farther from zero as actually observed?

In this example, the P value is less than 0.05 (and even less than 0.01), so we conclude that the null hypothesis is very unlikely to be true and that the observed linear relationship is unlikely to be caused by a coincidence of random sampling.

ASSUMPTIONS: LINEAR REGRESSION

Assumption: The model is correct

In many experiments, the relationship between X and Y is curved, making simple linear regression inappropriate. For example, linear regression of the data in Figure 23.3 resulted in a regression line that is nearly horizontal and a very high P value (0.9). X and Y are clearly related, just not linearly.

In the example, we know that the model cannot be accurate over a broad range of X values. At some values of X, the model even predicts that Y will be negative, a biological impossibility. But the linear regression model is useful within the range of X values actually observed in the experiment, so we only need to assume that the relationship between X and Y is linear within that range.

Assumption: The scatter of data around the line is Gaussian

Linear regression analysis assumes that the scatter of data around the model (the true best-fit line) is Gaussian. The confidence intervals and P values cannot be interpreted if the distribution of scatter is far from Gaussian or if some of the values are outliers from another distribution.

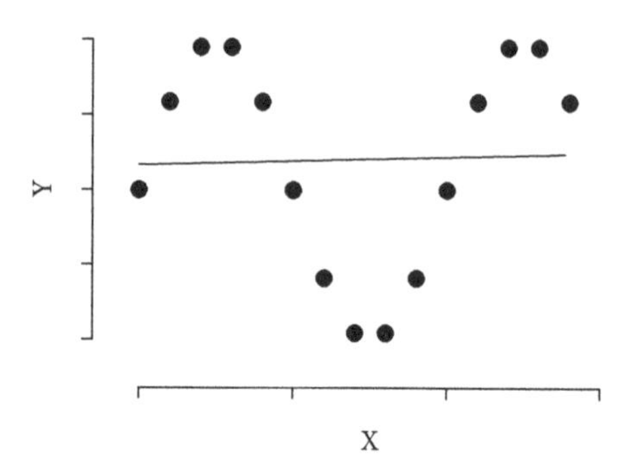

Figure 23.3. Linear regression only looks at linear relationships.
X and Y are definitely related here, just not in a linear fashion. The line drawn by linear regression misses the relationship entirely.

Assumption: The variability is the same everywhere

Linear regression assumes that the scatter of points around the best-fit line has the same standard deviation all along the curve. This assumption is violated if the points with high (or low) X values tend to be farther from the best-fit line. It is possible to bypass this assumption by differentially weighting the data points—in other words, giving more weight to the points with small variability and less weight to the points with lots of variability.

Assumption: Data points (residuals) are independent

In this example, the data are independent, because each point on the graph is from a different individual. If two of the individuals were twins (or even siblings), the assumption of independence would be questionable.

Assumption: The X and Y values are not intertwined

If the value of X is used to calculate the value of Y (or the value of Y is used to calculate the value of X), then linear regression calculations will be misleading. One example is a Scatchard plot, which is used by pharmacologists to summarize binding data. The Y value (drug bound to receptors divided by drug free in solution) is calculated from the X value (free drug), so linear regression is not appropriate. Another example would be a graph of midterm exam scores (X) versus total course grades (Y). Because the midterm exam score is a component of the total course grade, linear regression is not valid for these data.

Assumption: The X values are known precisely

Linear regression assumes that the true X values are known and that all the variation is in the Y direction. If X is something you measure (rather than a control) and the measurement is not precise, the linear regression calculations might be misleading.

COMPARISON OF LINEAR REGRESSION AND CORRELATION

The example data have now been analyzed twice, first using correlation (see Chapter 22) and then by linear regression. The two analyses are similar, yet distinct.

Correlation quantifies the degree to which two variables are related, but it does not fit a line to the data. The correlation coefficient tells you the extent to which one variable tends to change when the other one does as well as the direction of that change. The confidence interval of the correlation coefficient can only be interpreted when both X and Y are measured and both are assumed to follow Gaussian distributions. In the example, the experimenters measured both insulin sensitivity and %C20–22. You cannot interpret the confidence interval of the correlation coefficient if the experimenters manipulated (rather than measured) X.

With correlation, you don't have to think about cause and effect, and it doesn't matter which variable you call X and which you call Y. If you reversed the definition, all of the results would be identical. With regression, you do need to think about cause and effect. Linear regression finds the line that best predicts Y from X, and that line is not the same as the line that best predicts X from Y.

It made sense to interpret the linear regression line only because the investigators in our example hypothesized that the lipid content of the membranes would influence insulin sensitivity and therefore defined %C20–22 to be X and insulin sensitivity to be Y. The results of linear regression (but not correlation) would be different if the definitions of X and Y were swapped (so that changes in insulin sensitivity would somehow change the lipid content of the membranes).

With most data sets, it makes sense to calculate either linear regression or correlation, but not both. The example here (lipids and insulin sensitivity) is one for which both correlation and regression make sense. The values of R^2 and P are the same whether computed by a correlation or a linear regression program.

LINGO

Model

A *model* is an equation that describes the relationship between variables.

Parameters

The goal of linear regression is to find the values of the slope and intercept that make the line come as close as possible to the data. The slope and the intercept are called *parameters*.

Regression

Why the strange term *regression*? In the 19th century, Sir Francis Galton studied the relationship between parents and children. Children of tall parents tended to be shorter than their parents. Children of short parents tended to be taller than their parents. In each case, the height of the children reverted—or "regressed"—toward the mean height of all children. Somehow, though, the term *regression* has taken on a much more general meaning.

Residuals

The vertical distances of the points from the regression line are called *residuals.* A residual is the discrepancy between the actual Y value and the Y value predicted by the regression model.

Least squares

Linear regression finds the value of the slope and intercept that minimizes the sum of the squares of the vertical distances of the points from the line. This goal gives linear regression the alternative name *least squares*, or *linear least squares.* Not all regression is based on this principle, however. For example, logistic regression (which is briefly explained in Chapter 24) is not based on least squares.

Linear

The word *linear* has a special meaning to mathematical statisticians. It can be used to describe the mathematical relationship between model parameters and the outcome. Surprisingly, it is possible for the relationship between X and Y to be curved but for the mathematical model to be considered linear.

Independent and dependent variables

The linear regression model predicts an outcome, Y, from X. The Y variable is called the *dependent variable*, and the X variable is called the *independent variable.*

COMMON MISTAKES

Mistake: Concluding there is no relationship between X and Y when R^2 is low

A low R^2 value from linear regression means there is little *linear* relationship between X and Y. But not all relationships are linear! Linear regression of the data in Figure 23.3 reports $R^2 = 0.001$, but X and Y are clearly related (just not linearly).

Mistake: Fitting rolling averages or smoothed data

The left side of Figure 23.4 plots the number of hurricanes over time. These simulated data were chosen randomly, so there is no underlying trend. The right side of Figure 23.4 shows a *rolling average* (adapted from Briggs, 2008b). This is one way to create *smoothed data.* The smoothed graph shows a clear upward trend, the R^2 is much higher, and the P value is very low. But this trend is entirely an artifact of smoothing. Calculating the rolling average makes any random swing to a high (or a low) value look more consistent than it really is, because neighboring values also become high (or low). Accordingly, the regression line is misleading, and the P value and R^2 are meaningless.

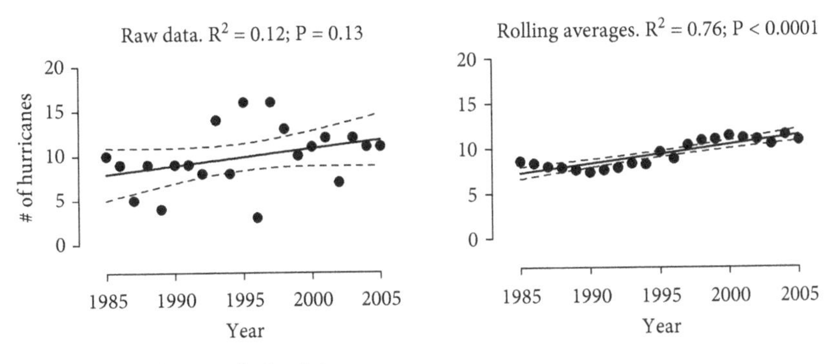

Figure 23.4. Don't smooth the data.
Smoothing, or computing a rolling average (which is a form of smoothing), appears to reduce the scatter of data and gives misleading results.

Mistake: Fitting data when X and Y are intertwined

When interpreting the results of linear regression, make sure that the X and Y axes represent separate measurements. If the X and Y values are intertwined (e.g., if Y is computed from X), the results will be misleading.

Mistake: Not thinking about which variable is X and which variable is Y

Unlike correlation, linear regression calculations are not symmetrical with respect to X and Y. Switching the labels X and Y will produce a different regression line (unless the data are perfect, with all points lying directly on the line). This makes sense, because the whole point is to find the line that best predicts Y from X, which is not the same as the line that best predicts X from Y.

Mistake: Looking at numerical results of regression without viewing a graph

Figure 23.5 shows four linear regressions designed by Anscombe (1973). The results of linear regression (slope, intercept, R^2, P value, and confidence intervals) are identical for all four data sets. But a glance at the graphs shows you that the data are very different!

Mistake: Extrapolating beyond the data

The top left graph in Figure 23.6 shows data that are fit by a linear regression model very well. The data range from X = 1 to X = 15. The top right and bottom graphs predict what will happen at later times. The right graph on the top shows the prediction of linear regression. The bottom two graphs show predictions of two other models. The three models (top right, bottom) make very different predictions at late time points. Which prediction is most likely to be correct? It is impossible to answer that question without more data or at least some information and theory about the scientific context. Figure 23.7 points out the folly of extrapolating a linear regression line

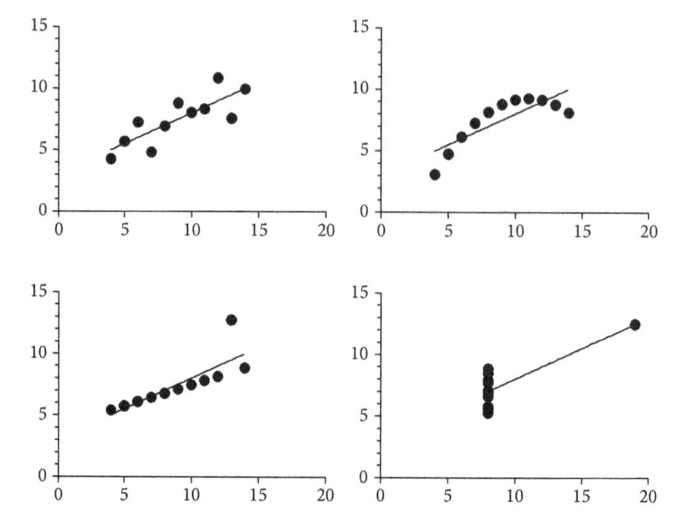

Figure 23.5. Look at a graph as well as at the numerical results.

The linear regressions of these four data sets all have identical best-fit values for slope, best-fit values for the Y intercept, and R^2 values. Yet the data are far from identical.

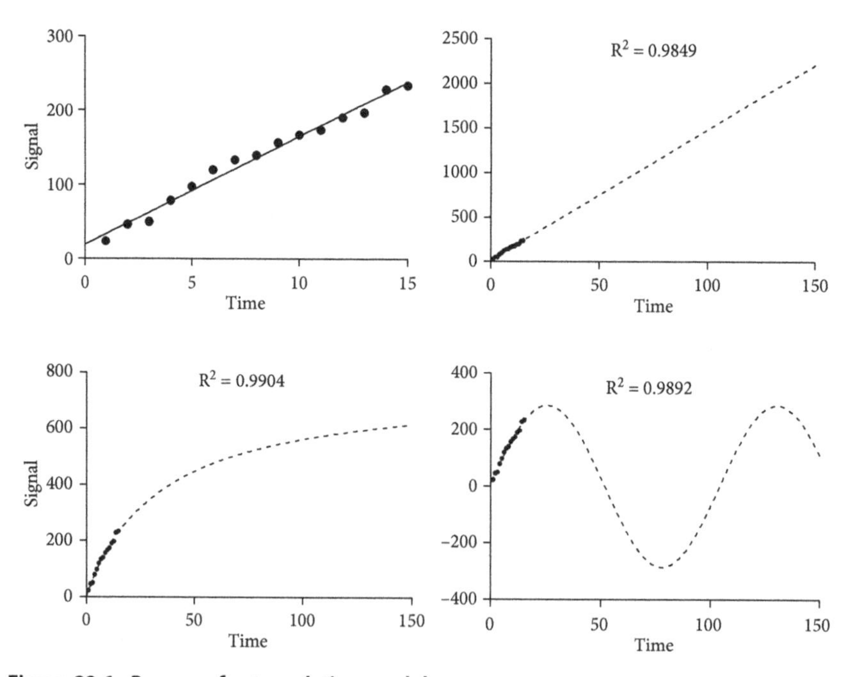

Figure 23.6. Beware of extrapolating models.

(Top left) Data that fit a linear regression model very well. (Top right) The prediction of linear regression at later time points. (Bottom) Predictions of two other models that actually fit the data slightly better than a straight line does. Predictions of linear regression that extend far beyond the collected data may be very wrong.

Figure 23.7. Beware of extrapolation.

Extrapolating can lead to silly conclusions!

Source: xkcd.com.

Mistake: Over interpreting a small P value

Figure 23.8 shows simulated data with 500 points fit by linear regression. The P value is only 0.0302, so the deviation of the line from a horizontal line is statistically significant. But look at R^2, which is only 0.0094. The linear

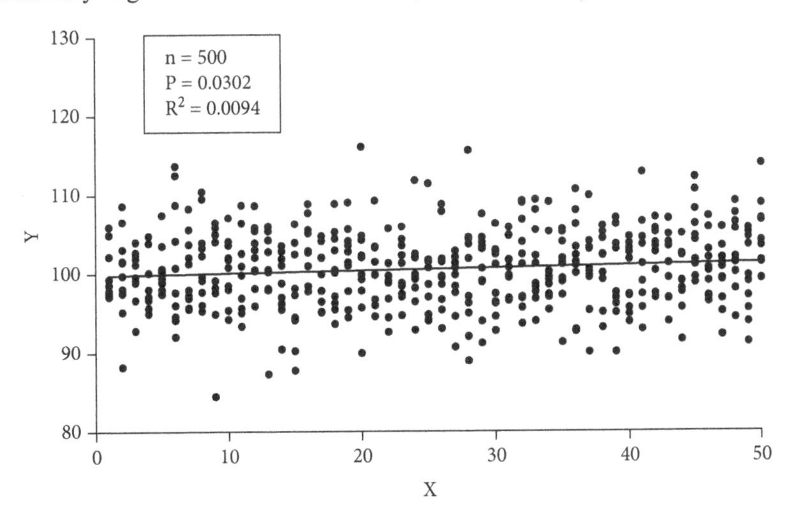

Figure 23.8. Beware of small P values from large samples.

The line was fit by linear regression. It is statistically different from horizontal ($P = 0.0302$). But R^2 is only 0.0094, so the regression model explains less than 1% of the variance. The P value is small, so these data would not be expected if the true slope were exactly horizontal. However, the deviation from horizontal is trivial. You'd be misled if you focused on the P value and the conclusion that the slope differs significantly from zero.

regression model accounts for less than 1% of the overall variation. The low P value should convince you that the true slope is unlikely to be horizontal. However, the discrepancy is tiny. There may possibly be some scientific field in which such a tiny effect is considered important or worthy of follow up. But in most fields, this kind of effect is completely trivial even though it is statistically significant. Focusing on the P value can lead you to misunderstand the findings.

Mistake: Saying that you are fitting data to a model

Regression finds values of the parameters that best fit a model to the data. It is incorrect to say that you are fitting the data to a model, as that implies you are changing the values of the data.

Q & A

Do the X and Y values need to have the same units to perform linear regression?

No, but they can. In the insulin example, X and Y are in different units.

Can linear regression work when all X values are the same or all Y values are the same?

No. The whole point of linear regression is to predict Y based on X. If all the X values are the same, they won't help predict Y. If all the Y values are the same, there is nothing to predict.

Is the standard error of the slope the same as the standard error of the mean?

No. The standard error is a way to express the precision of a computed value (parameter). The first standard error encountered in this book happened to be the standard error of the mean (see Chapter 10). The standard error of a slope is quite different. Standard errors can also be computed for many other parameters.

Will the regression line be the same if you exchange X and Y?

Linear regression fits a model that best predicts Y from X. If you swap the definitions of X and Y, the regression line will be different unless the data points line up perfectly so that every point is on the line. However, swapping X and Y will not change the value of R^2.

If you analyze the same data with linear regression and correlation (see Chapter 22), how do the results compare?

If you square the correlation coefficient, the value will equal R^2 from linear regression. The P value testing the null hypothesis that the population correlation coefficient is zero will match the P value testing the null hypothesis that the population slope is zero.

R^2 or r^2?

Both forms are used with linear regression, and there is no distinction.

CHAPTER SUMMARY

- Linear regression fits a model to the data to determine the value of the slope and intercept that makes a line best fit that data.
- Linear regression also computes the 95% CI for the slope and intercept and can plot a confidence band for the regression line.
- Goodness of fit is quantified by R^2.
- Linear regression reports a P value, which tests the null hypothesis that the population slope is horizontal.

CHAPTER 24

Nonlinear, Multiple, and Logistic Regression

NONLINEAR REGRESSION

Just as linear regression fits a line through your data, *nonlinear regression* fits a curve. It is a form of *curve fitting*. The calculations of nonlinear regression are far more complicated than the calculations of linear regression, and I think this is the reason nonlinear regression is omitted from most statistics texts. But nonlinear regression is widely used in many scientific fields and is easy to understand the basics.

The first step in nonlinear regression is choosing a model. In most cases, scientists can choose a standard model explained in texts and built into nonlinear regression programs. The model is an equation that defines Y as a function of X and one or more parameters. In linear regression, the parameters are the slope and intercept. In nonlinear regression, the parameters might include dissociation constants, rate constants or almost anything else. In addition to choosing a model, someone using nonlinear regression software also needs to specify which parameters should be fit to the data and which (if any) should be fixed to constant values.

Nonlinear regression depends on the same set of assumptions listed in the previous chapter for linear regression. Note in particular the first assumption, that the model is correct. Nonlinear regression doesn't choose the best model for your data. Rather, you assume the model is correct and find the best-fit values of the parameters. You can also compare the fits of two (or more) alternative models.

Interpreting results from nonlinear regression is not all that different than interpreting results from linear regression. Look at the graph with the curve superimposed on the data, and at the parameter values with their confidence intervals. How you interpret those parameters depends on your understanding of how the chosen model expresses a physiological (or chemical, or genetic, etc.) model and how that relates to the scientific goals of the experiment.

MULITPLE AND LOGISTIC REGRESSION

The need for regression with multiple independent variables

In laboratory experiments, you can generally control all the variables. You change one variable, measure another, and then analyze the data with a standard statistical test.

But in some kinds of experiments and many observational studies, you must analyze how one variable (the *dependent variable*) is influenced by several variables (called *independent variables*). Scientists using such methods usually have one of these goals:

- To assess the impact of one variable after accounting for others. Does a drug work after accounting for age differences between the patients who received the drug and those who received a placebo? Does an environmental exposure increase the risk of a disease after taking into account other differences between people who were and were not exposed to that risk factor?
- To create an equation for making useful predictions. Given the data we know now, what is the chance that this particular man with chest pain is having a myocardial infarction (heart attack)? Given several variables that can be measured easily, what is the predicted cardiac output of this patient?
- To understand scientifically how changes in each of several variables contribute to explaining an outcome of interest.

Interpreting results from multiple regression and logistic regression

The key results are the best-fit values for each parameter, along with the corresponding confidence intervals. When the outcome (dependent variable) is continuous, the corresponding parameter is the amount by which the Y variable goes up when the corresponding X variable is incremented by 1.0 (and all other X variables are held constant). When the X variable is binary, then the corresponding parameter is the average difference in Y comparing one value of X (e.g., male) versus the other (e.g., female). When the outcome is binary (logistic regression), the easiest way to understand the parameters is to transform them to odds ratio, but you'll need to read elsewhere to learn about this.

Regression programs often report a P value for each parameter testing the null hypothesis that the true population value of that parameter is zero. Why zero? When a regression coefficient (parameter) equals zero, then the corresponding independent variable has no effect in the model, so the P value is used to test whether a particular parameter makes a statistically significant contribution to the outcome.

LINGO

The terminology is a bit confusing.

If the outcome (the dependent variable Y) is binary, the method used to fit the model is called *logistic regression*. If the outcome is continuous, the method used to fit the model is called *linear* (or *nonlinear*) *regression*.

If there is only one independent (X) variable, regression is prefixed with *simple*, while if there are more than one independent variables, the regression is prefixed with *multiple*. The linear regression explained in Chapter 23 has one independent (X) variable and so is *simple linear regression*, but the word "simple" is often omitted. When a model has more than one independent variables, the prefix *multiple* is used. With a continuous dependent variable, the method is *multiple linear regression*, but the word "linear" is often omitted, and you'll see mentions of "multiple regression". When the outcome is binary, models almost always include more than one independent variable so the prefix *multiple* is often omitted.

Beware of the term *multivariate*, which is used inconsistently. Mostly it is used to describe methods that analyze data with more than one outcome. Sometimes it is used to describe models with more than one independent variable, which are more properly called *multivariable* models.

COMMON MISTAKES

Mistake: Avoiding nonlinear regression because you think it is too complicated

The mathematical foundations of nonlinear regression are complicated. But using a nonlinear regression program and interpreting the results is just a tiny bit harder than using linear regression.

Mistake: Transforming curved data in order to use linear regression

Before nonlinear regression became readily available, scientists often transformed their data to make the graph linear and used linear regression. Nonlinear regression is easier and gives more accurate results.

Mistake: Having too many independent variables

The goal of regression, as in all of statistics, is to analyze data from a sample and make valid inferences about the overall population. That goal cannot always be met using multiple regression techniques. If you have too many independent variables, you are likely to make conclusions that apply to the sample data but are not really true in the population. When the study is repeated, the conclusions will not be reproducible. This problem is called *overfitting* (Babyak, 2004).

How many independent variables is too many? It depends on sample size and the scientific goals of the study. For multiple regression, one rule of thumb is that you need to have at least 10 to 20 subjects per independent variable. Fitting a model with five independent variables thus requires at least 50 to 100 subjects. To use that rule with logistic regression, count the subjects with the rarer of the two outcomes. Let's say that outcome is the incidence of infection after surgery, which you expect to happen to 10% of the patients. If the model has 5 independent variables and you are using that same rule of thumb, you need enough subjects so at least 50 to 100 will be infected, which means you need at least 500 to 1000 subjects in the study.

Mistake: Selecting independent variables automatically

Some multiple and logistic regression programs automatically choose independent variables. You just put all the data into the program, make a few general choices, and the program creates a simpler model using only a subset of the possible independent variables. The problem with this approach is multiple comparisons. For example, if the investigator starts with 20 possible independent variables, automatic variable selection essentially compares the fits of more than a million models.

Q & A

If an investigator used variable selection, how can I interpret the results?

The consequences of variable selection are pervasive and serious (Flom & Cassell, 2007; Harrell, 2001):

- The final model fits the experimental data much better than it will fit future data.
- The parameters for the variables that get selected to stay in the model are "too high" (too far from zero). If you repeat with new data, those parameters are likely to be closer to zero.
- The confidence intervals are too narrow, so you think you know the parameter values with more precision than is warranted.
- When you test whether the parameters are statistically significant, the P values are too low.

Be very wary when interpreting multiple (or logistic) regression results where the reported model uses only a subset of the independent variables the investigator started with.

Are these problems only with *automatic* variable selection?

No. The problems are the same whether the program selected variables automatically, the investigator manually compared several models, or the results of some simpler screening analysis (usually correlation) decided which variables were included in the model.

What is collinearity?

If two (or more) independent X variables are highly correlated, the results of multiple or logistic regression may be misleading. The multiple regression calculations assess the *additional* contribution of each independent variable after accounting for all the other independent variables. When independent variables are correlated, each variable makes little additional contribution. This is called *collinearity*. For example, if you include both weight and height into a model, the two variables are correlated (people who are taller also tend to be heavier), so the confidence intervals for the corresponding regression coefficients will be too wide, and the corresponding P values will be too high. You may conclude that neither height nor weight influences the outcome variable, and that conclusion might be wrong.

CHAPTER SUMMARY

- Like linear regression, nonlinear regression fits a model to your data to determine the best-fit values of parameters.
- Using a nonlinear regression program is not much harder than using a linear regression program.
- Regression with multiple independent variables is used to assess the impact of one variable after correcting for the influences of others, to predict outcomes from several variables, or to try to tease apart complicated relationships among variables.
- Multiple linear regression is used when the outcome (Y) variable is continuous. Logistic regression is used when the outcome is binary.
- Be wary of the term *multivariate*, which is used inconsistently.
- Selecting a subset of independent variables to make a simpler model is appealing, but the results can be misleading. It is a form of multiple comparisons.
- If you have too many independent variables, you will encounter overfitting. The results will not be reproducible.

CHAPTER 25

Common Mistakes to Avoid When Interpreting Published Statistics

Most of the chapters in this book ended with a Common Mistakes section. This final chapter describes general mistakes that don't really fit into any particular chapter.

MISTAKE: NOT RECOGNIZING PUBLICATION BIAS

By the time you read a paper, a great deal of selection has occurred. When experiments are successful, scientists tend to continue the project, while less successful projects get abandoned. When the project is done, scientists are more likely to write up projects that lead to remarkable results or keep analyzing the data in various ways to extract a statistically significant conclusion. Finally, journals are more likely to publish "positive" studies. If the null hypothesis were true, you would expect a statistically significant result in 5% of experiments. But those 5% are more likely to get published than the other 95%. This is called *publication bias*.

If many studies were performed, you might expect some studies to find larger effects, some studies to find smaller effects, and for the average effect size to be close to the truth. However, studies with small effects tend not to get published. On average, therefore, the studies that do get published tend to report effect sizes that overestimate the true effect (Ioannidis, 2008).

MISTAKE: TESTING HYPOTHESES SUGGESTED BY THE DATA

The trap occurs when a scientist studies many variables and many subgroups, discovers an intriguing relationship, and then publishes the results so that it appears the hypothesis was stated before the data collection began. This is called *HARKing* (short for *Hypothesizing After the Results are Known*; Kerr, 1998), or *double dipping* (Kriegeskorte et al., 2009). An XKCD cartoon points out the folly of this approach (Figure 25.1).

Figure 25.1. Nonsense conclusions via HARKing (Hypothesizing After the Results are Known).

The conclusion about green jelly beans was one of twenty comparisons. It is not fair to report on the one "statistically significant" conclusion without reporting all the findings (see Chapter 17).

Source: xkcd.com.

MISTAKE: MAKING A CONCLUSION ABOUT CAUSATION WHEN THE DATA ONLY SHOW CORRELATION

Messerli (2012) graphed the total number of Nobel Prizes ever won by citizens of each country as a function of chocolate consumption (Figure 25.2). Both variables are standardized to the country's population. The correlation is amazingly strong, with r = 0.79. A P value testing the null hypothesis of no real correlation is tiny.

Of course, these data don't prove that eating chocolate helps people win a Nobel Prize, or that increasing chocolate imports will increase the number of Nobel Prizes the residents of that country will win. Many variables differ among countries, and some of those probably correlate with both chocolate consumption and number of Nobel Prizes.

The cartoons in Figure 25.3 drive home the point that correlation does not prove causation.

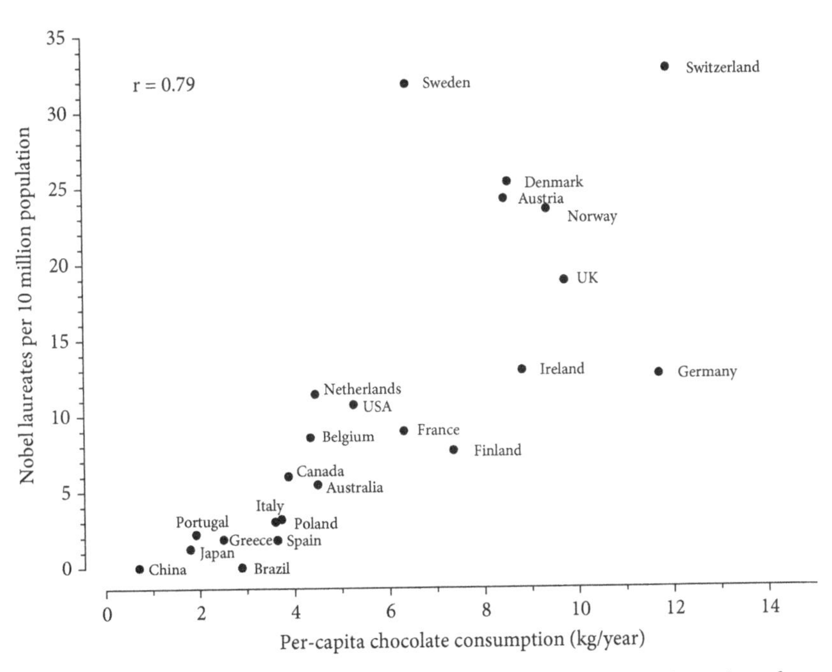

Figure 25.2. Correlation between average chocolate consumption and number of Nobel Prize winners by country.

This figure is redrawn from Messerili (2012). The Y axis plots the total number of Nobel Prizes won by citizens of each country. The X axis plots chocolate consumption in a recent year (different years for different countries, based on the availability of data). Both X and Y values are normalized to the country's current population. The correlation is amazingly strong, but this doesn't prove that eating chocolate will help you win a Nobel Prize.

Figure 25.3. Correlation does not imply causation.

Source: (Top) xkcd.com. (Bottom) DILBERT © 2011 Scott Adams. Used By permission of UNIVERSAL UCLICK. All rights reserved.

MISTAKE: OVER INTERPRETING STUDIES THAT MEASURE A PROXY OR SURROGATE OUTCOME

For many years, people who had myocardial infarctions (heart attacks) were treated with antiarrhythmic drugs that prevent extra heartbeats (premature ventricular contractions) and thus were thought to reduce the incidence of sudden death due to arrhythmia. The logic was clear. Abnormal results (extra beats) on an electrocardiogram were known to be associated with sudden death, and treatment with antiarrhythmic drugs was known to reduce the number of extra beats. So it made sense to conclude that taking those drugs would extend life. The evidence was compelling enough that the FDA even approved use of these drugs for this purpose. But a randomized study to directly test the hypothesis that antiarrhythmic drugs would reduce sudden death showed just the opposite. Patients taking two specific antiarrhythmic drugs had fewer extra beats (the *proxy* or *surrogate* variable) but were more likely to die (Cardiac Arrhythmia Suppression Trial [CAST] Investigators, 1989). Fisher and Van Belle (1993) summarize the background and results of this trial.

Another example is the attempt to prevent heart attacks by using drugs to raise high-density lipoprotein (HDL) levels. Low levels of HDL ("good cholesterol") are associated with an increased risk of atherosclerosis and heart disease.

Pfizer Corporation developed a drug that elevates HDL, with great hope that it would prevent heart disease. Barter and colleagues (2007) gave the drug to thousands of patients with a high risk of cardiovascular disease. Low-density lipoprotein (LDL, or "bad cholesterol") decreased 25%, and "good" (HDL) cholesterol increased 72%. The confidence intervals were narrow, and the P values were tiny. If the goal were to improve cholesterol levels, the drug was a huge success. Unfortunately, however, treatment with that drug also increased the number of heart attacks by 21% and the number of deaths by 58%. Had you only seen the changes in the proxy variables (cholesterol levels) you would have been mislead about the effectiveness of the drugs.

MISTAKE: OVER INTERPRETING DATA FROM AN OBSERVATIONAL STUDY

Munger and colleagues (2013) wondered whether deficiency of vitamin D predisposed people to developing type I diabetes. The researchers compared a group of people with diabetes to a group of people without diabetes who were similar in other ways and then compared the vitamin D levels in blood samples drawn before disease onset. They found that people with average levels greater than 100 nmol/L had a about half the risk of developing diabetes than those whose average levels were less than 75 nmol/L. The risk ratio was 0.56, with a 95% CI ranging from 0.35 to 0.90 (P = 0.03).

Do these findings mean that taking vitamin D supplements will prevent diabetes? No. The association of low vitamin D levels and onset of diabetes could be explained in many ways. Sun exposure increases vitamin D levels. Perhaps exposure to sunlight also creates other hormones (which we may not have yet identified) that decrease the risk of diabetes. Perhaps the people who are exposed more to the sun (and thus have higher vitamin D levels) also exercise more, and it is the exercise that helps prevent diabetes. Perhaps the people with higher vitamin D levels drink more fortified milk, and it is the calcium in the milk that helps prevent diabetes. The only way to find out for sure whether ingestion of vitamin D prevents diabetes is to conduct an experiment comparing people who are given vitamin D supplements with those who are not.

Although observational studies are much easier to conduct than experiments, data from experiments are more definitive. Observational studies often require more complicated analyses and yield less certain results. To emphasize this point, Spector and Vesell (2006) reviewed five hypotheses suggested by observational studies that turned out not to be valid when tested with clinical experiments (see Table 25.1).

MISTAKE: BEING FOOLED BY REGRESSION TO THE MEAN

Imagine that you and your classmates are given a true/false exam in a subject you know nothing about and that the test is written in a language none of you know.

		RESULTS FROM . . .	
INTERVENTION	**INCIDENCE OF**	**OBSERVATIONAL STUDIES**	**EXPERIMENT**
Hormone replacement therapy after menopause	Cardiovascular events (myocardial infarction, sudden death and strokes)	Decrease	Increase
Megadose vitamin E	Cardiovascular events	Decrease	No change
Low-fat diet	Cardiovascular events and cancer	Decrease	No change
Calcium supplementation	Fractures and cancer	Decrease	No change
Vitamins to reduce homocysteine	Cardiovascular events	Decrease	No change

Table 25.1. Five hypotheses suggested by observational studies proven not to be true by experiment (from Spector and Vesell, 2006).

All that you and your classmates can do is guess at each answer. On average, you expect to answer correctly about half the time. Some students, by chance, will do better than average, and some will do worse.

Now, imagine that you focus on the people who got the worst scores on the test and intervene to help them do better next time. Maybe you give them nutritional supplements, or perhaps you coach them in exercise or meditation. Then they take a second, similar test. Since they still know nothing about the subject matter and can't even understand the language, all they can do is guess. On average, their score will be 50%. That means that on average, these students will do better on the second exam! But it would be a mistake to attribute the improvement in the score to the intervention you arranged between the two tests. Instead, it is purely a matter of chance. If you selected the students who did the best on the first exam, you'd find that on average, they would do worse on the second exam.

This is an example of *regression to the mean*. If you had attributed the increase in exam scores to the treatment, your mistake would be called a *regression fallacy*. More examples: People who are especially lucky at picking stocks one year are likely to be less lucky the next year. An athlete who does extremely well in one season is likely to perform more poorly the next season. People who have extremely high blood pressure on one exam are likely to have a lower blood pressure on a repeat exam regardless of treatment. Someone who comes to a clinic because of severe headaches is likely to have fewer, or less severe, headaches the next week regardless of treatment.

CHAPTER 26

Review

THE FUNDAMENTAL IDEAS OF STATISTICS

Statistical inference helps you make general conclusions from limited data, so conclusions are always presented in terms of probability
Be wary if you ever encounter statistical conclusions that seem 100% definitive.

All statistical tests are based on assumptions
Review the list of assumptions before interpreting any statistical results.

Decisions about how to analyze data should have been made in advance
Otherwise the investigators may be P-hacking.

Statistics is only part of interpreting data
Also think about study design and experimental methods.

Many statistical terms are also ordinary words
Don't mistakenly give a statistical term an ordinary meaning.

The standard error of the mean does not quantify variability
The standard deviation and standard error of the mean are often confused.

Confidence intervals quantify precision
All values (means, difference, ratio, etc.) computed from data should be reported with a confidence interval.

Every P value tests a null hypothesis
You cannot understand a P value until you can precisely state the corresponding null hypothesis.

The concept of statistical significance is designed to help you make a decision based on one result
If you don't plan to use this one result to make a crisp decision, the concept of statistical significance is not necessary.

"Statistically significant" does not mean the effect is large or scientifically important
It only means a difference (or association, or correlation) this large as or larger will happen less than 5% of the time (or some other stated value) by chance alone.

"Not significantly different" does not mean the effect is absent, small, or scientifically irrelevant
All you can conclude that the observed results are not inconsistent with the null hypothesis.

The term *significant* has two meanings, so is often misunderstood
Avoid the term when possible.

Multiple comparisons make it hard to interpret statistical results
To correctly interpret statistical analyses, all analyses must be planned before collecting data, and all planned analyses must be conducted and reported.

STATISTICAL VOCABULARY BY CHAPTER

2. The Complexities of Probability

- Bayesian statistics
- Gambler's fallacy
- Model
- Odds
- Population
- Probability
- Sample
- Statistics
- Subjective probability

3. From Sample to Population

- Bias
- Convenience sample
- Model
- Parameters
- Population
- Random sample
- Sample
- Sampling error
- Selection bias
- Systematic sample

4. Confidence Intervals

- Biased
- Bayesian statistics
- Binomial distribution
- Confidence interval (CI)
- Confidence level
- Confidence limit
- Credible interval
- Cumulative binomial distribution
- Descriptive statistics
- Estimate
- Inferential statistics
- Interval estimate
- Point estimate
- Random sample
- Sampling error

5. Types of Variables

- Binary variable
- Continuous variable
- Dummy variable
- Interval variable
- Nominal variable
- Ordinal variable
- Ratio variable

6. Graphing Variability

- Bias
- Box-and-whiskers plot
- Column scatter plot
- Continuous data
- Dot plot
- Error
- Experimental error
- Frequency distribution histogram
- Mean
- Median
- Moving average

- Precision
- Rolling average
- Smoothed data

7. Quantifying Variation

- Coefficient of variation (CV)
- Five-number summary
- Interquartile range
- Mean
- Median
- Percentiles
- Population standard deviation
- Quartiles
- Range
- Sample standard deviation
- Standard deviation (SD)
- Variance

8. The Gaussian Distribution

- Central limit theorem
- Gaussian distribution
- Normal distribution
- Standard normal distribution
- z

9. The Lognormal Distribution and Geometric Mean

- Antilogarithm
- Arithmetic mean
- Common logarithms
- Geometric mean
- Logarithm
- Lognormal distribution

10. Confidence Interval for a Mean

- Confidence limit
- Sampling error
- Standard error of the mean (SEM)

11. Error Bars

- Dynamite plot
- Error bar

12. Comparing Groups with Confidence Intervals

- Confounding variables
- Metaanalysis
- Risk ratio
- Relative risk

13. Comparing Groups with P Values

- Fisher's exact test
- Log-rank test
- Null hypothesis
- P value
- Unpaired t test

14. Statistical Significance and Hypothesis Testing

- Alpha (α)
- Five-sigma threshold
- P-hacking
- Significance level
- Stargazing
- Statistical hypothesis testing
- Statistically significant
- Type I error
- Type II error
- Type III error
- Type S error

15. Interpreting a Result That Is (Or Is Not) Statistically Significant

- False discovery
- Statistically significant
- Type I error
- Type II error
- Type III error
- Type S error

16. How Common Are Type I Errors?

- Bayesian approach
- False discovery rate (FDR)
- Prior probability

17. Multiple Comparisons

- Bonferroni correction
- Family of comparisons
- Familywise error rate
- Genome-wide association studies

- Multiple comparisons
- Per-comparison error rate
- Per-experiment error rate
- Primary outcome
- Secondary outcome

18. Statistical Power and Sample Size

- Ad hoc sequential sample size determination
- Effect
- Power
- Power analysis
- Standard effect sizes
- Underpowered

19. Commonly Used Statistical Tests

- Attributable risk
- Chi-square test
- Dunnett's test
- Fisher exact test
- Kaplan-Meier graphs
- Log-rank test
- Mann-Whitney test
- Nonparametric tests
- One-way analysis of variance (ANOVA)
- One factor ANOVA
- Paired t test
- Parametric tests
- Relative risk
- Tukey test
- Two-sample t test
- Unpaired t test
- Wilcoxon matched pairs test
- Wilcoxon signed rank test

20. Normality Tests

- Gaussian distribution
- Normal distribution
- Normality test

21. Outliers

- Outlier
- Outlier test

22. Correlation

- Correlation
- Correlation coefficient (r)
- Coefficient of determination (r^2)
- Covariation
- Pearson correlation coefficient
- r
- R^2
- Spearman correlation coefficient

23. Simple Linear Regression

- Confidence bands
- Dependent variable
- Independent variable
- Intercept
- Least squares
- Linear
- Linear regression
- Linear least squares
- Model
- Parameter
- Regression
- Residuals
- Rolling average
- Slope
- Smoothed data

24. Nonlinear, Multiple, and Logistic Regression

- Curve fitting
- Dependent variable
- Independent variable
- Logistic regression
- Multiple regression
- Multivariate
- Nonlinear regression
- Overfitting
- Parameter

25. Common Mistakes

- HARKing
- Proxy variable
- Publication bias
- Regression fallacy
- Regression to the mean

REFERENCES

Agnelli, G., Buller, H. R., Cohen, A., Curto, M., Gallus, A. S., Johnson, M., Porcari, A., et al. (2013). Apixaban for extended treatment of venous thromboembolism. *New England Journal of Medicine, 368*, 699–708. doi:10.1056/NEJMoa1207541

Alere. (2014). DoubleCheckGold™ HIV 1&2. Retrieved May 14, 2014, from http://www.alere.com/ww/en/product-details/doublecheckgold-hiv-1-2.html

Allen, M. C., Donohue, P. K., & Dusman, A. E. (1993). The limit of viability—Neonatal outcome of infants born at 22 to 25 weeks' gestation. *New England Journal of Medicine, 329*, 1597–1601.

Altman, D. G., & Bland, J. M. (1995). Absence of evidence is not evidence of absence. *BMJ (Clinical Research Ed.), 311*, 485.

———. (1998). Time to event (survival) data. *BMJ (Clinical Research Ed.), 317*, 468–469.

Anscombe, F. J. (1973). Graphs in statistical analysis. *The American Statistician, 27*, 17–21.

Austin, P. C., & Goldwasser, M. A. (2008). Pisces did not have increased heart failure: Data-driven comparisons of binary proportions between levels of a categorical variable can result in incorrect statistical significance levels. *Journal of Clinical Epidemiology, 61*, 295–300.

Austin, P. C., Mamdani, M. M., Juurlink, D. N., & Hux, J. E. (2006). Testing multiple statistical hypotheses resulted in spurious associations: A study of astrological signs and health. *Journal of Clinical Epidemiology, 59*, 964–969.

Babyak, M. A. (2004). What you see may not be what you get: A brief, nontechnical introduction to overfitting in regression-type models. *Psychosomatic Medicine, 66*, 411–421.

Barter, P. J., Caulfield, M., Eriksson, M., Grundy, S. M., Kastelein, J. J., Komajda, M., Lopez-Sendon, J., et al. (2007). Effects of torcetrapib in patients at high risk for coronary events. *New England Journal of Medicine, 357*, 2109–2122.

Bausell, R. B. (2007). *Snake oil science: The truth about complementary and alternative medicine*. Oxford, UK: Oxford University Press. ISBN: 0195313682.

Begley, C. G. C., & Ellis, L. M. L. (2012). Drug development: Raise standards for preclinical cancer research. *Nature*, 483, 531–533.

Bennett, C. M., Baird, A. A., Miller, M. B., & Wolford, G. L. (2011). Neural correlates of interspecies perspective taking in the post-mortem Atlantic salmon: An argument for proper multiple comparisons correction. *Journal of Serendipitous and Unexpected Results, 1*, 1–5.

Bickel, P. J., Hammel, E. A., & O'Connell, J. W. (1975). Sex bias in graduate admissions: Data from Berkeley. *Science, 187*, 398–404.

Boos, D. D., & Stefanski, L. A. (2011). P-value precision and reproducibility. *The American Statistician, 65*, 213–221.

Borkman, M., Storlien, L. H., Pan, D. A., Jenkins, A. B., Chisholm, D. J., & Campbell, L. V. (1993). The relation between insulin sensitivity and the fatty-acid composition of skeletal-muscle phospholipids. *New England Journal of Medicine, 328*, 238–244.

Briggs, W. M. (2008a, July 25). On the difference between mathematical ability between boys and girls. Retrieved June 21, 2009, from http://wmbriggs.com/blog/?p=163/

———. (2008b, February 14). Do not calculate correlations after smoothing data. Retrieved June 21, 2009, from http://wmbriggs.com/blog/?p=86/

Cardiac Arrhythmia Suppression Trial (CAST) Investigators. (1989). Preliminary report: Effect of encainide and flecainide on mortality in a randomized trial of arrhythmia suppression after myocardial infarction. *New England Journal of Medicine, 321*, 406–412.

Central Intelligence Agency (2012), *The World Factbook*, viewed August 26, 2012, <httpscox://www.cia.gov/library/publications/the-world-factbook/fields/2018.html>.

Chan, A. W., Hrobjartsson, A., Haahr, M. T., Gotzsche, P. C., & Altman, D. G. (2004). Empirical evidence for selective reporting of outcomes in randomized trials: Comparison of protocols to published articles. *Journal of the American Medical Association, 291*, 2457–2465.

Cohen, J. (1988). *Statistical power analysis for the behavioral sciences,* 2nd ed. Hillsdale, NJ: Erlbaum. ISBN: 0805802835.

Colquhoun, D. (2003). Challenging the tyranny of impact factors. *Nature, 423*, 479–480.

———. (2013, April 13). Another update. Red meat doesn't kill you, but the spin is fascinating. Retrieved January 10, 2014, from http://www.dcscience.net/?p=5935

———. (2014). An investigation of the false discovery rate and the misinterpretation of P values. *R. Soc. Open Sci.* 1: 140216. http://dx.doi.org/10.1098/rsos.140216.

Cumming, G. (2008). Replication and p intervals: P values predict the future only vaguely, but confidence intervals do much better. *Perspectives on Psychological Science, 3,* 286–300.

———. (2011). *Understanding the new statistics: Effect sizes, confidence intervals, and meta-analysis.* New York: Routledge. ISBN: 978-0415879682.

Federal Election Commission. (2013, January 17). Official 2012 presidential general election results. Retrieved June 2, 2013, from http://www.fec.gov/pubrec/fe2012/2012presgeresults.pdf

Fisher, L. D., & Van Belle, G. (1993). *Biostatistics. A methodology for the health sciences,* New York: Wiley Interscience. ISBN: 0471584657.

Flom, P. L., & Cassell, D. L. (2007). Stopping stepwise: Why stepwise and similar selection methods are bad, and what you should use. NorthEast SAS Users Group. Retrieved September 17, 2012, from http://www.nesug.org/proceedings/nesug07/sa/sa07.pdf

Freedman, D. (2007). *Statistics,* 4th ed. New York: Norton. ISBN: 978-0393929720.

Gallup. (2012, September 18). Romney has support among lowest income voters. Retrieved Dec. 17, 2014 from http://www.gallup.com/poll/157508/romney-support-among-lowest-income-voters.aspx

Gelman, A. (1998). Some class-participation demonstrations for decision theory and Bayesian statistics. *The American Statistician, 52*, 167–174.

Gelman, A., & Feller, A. (2012, November 12). Red versus blue in a new light. *The New York Times*. Retrieved January 31, 2013, from http://campaignstops.blogs.nytimes.com/2012/11/12/red-versus-blue-in-a-new-light/

Gelman, A., & Stern, H. (2006). The difference between "significant" and "not significant" is not itself statistically significant. *The American Statistician, 60*, 328–331.

Gelman, A., & Tuerlinckx, F. (2000). Type S error rates for classical and Bayesian single and multiple comparison procedures. *Computational Statistics, 15*, 373–390.

Goddard, S. (2008, May 2). Is the earth getting warmer, or cooler? *The Register*. Retrieved June 13, 2008, from http://www.theregister.co.uk/2008/05/02/a_tale_of_two_thermometers/

Goodman, S. (2008). A dirty dozen: Twelve p-value misconceptions. *Seminars in Hematology, 45*, 135–140.

Gotzsche, P. C. (2006). Believability of relative risks and odds ratios in abstracts: Cross sectional study. *BMJ (Clinical Research Ed.), 333*, 231–234.

Gould, S. J. (1997). *Full house: The spread of excellence from Plato to Darwin*. New York: Three Rivers Press. ISBN: 0609801406.

Hankins, M. (2013, April 12). Still not significant. Retrieved June 28, 2013, from http://mchankins.wordpress.com/2013/04/21/still-not-significant-2/

Harrell, F. E. (2001). Regression Modeling Strategies: With Applications to Linear Models, Logistic Regression, and Survival Analysis. New York: Springer. ISBN: 978-0387952321.

Harter, H. L. (1984). Another look at plotting positions. *Communications in Statistics—Theory and Methods, 13*, 1613–1633.

Hartung, J. (2005). Statistics: When to suspect a false negative inference. In *American Society of Anesthesiology 56th annual meeting refresher course lectures*, Lecture 377, 1–7. Philadelphia: Lippincott.

Henderson, B. (2005). Open letter to Kansas School Board. Retrieved December 8, 2012, from http://www.venganza.org/about/open-letter/

Hsu, J. (1996). *Multiple comparisons: Theory and methods*. Boca Raton, FL: Chapman & Hall/CRC. ISBN: 0412982811.

Hviid, A., Melbye, M., & Pasternak, B. (2013). Use of selective serotonin reuptake inhibitors during pregnancy and risk of autism. *New England Journal of Medicine, 369*, 2406–2415.

Hyde, J. S., Lindberg, S. M., Linn, M. C., Ellis, A. B., & Williams, C. C. (2008). Diversity. Gender similarities characterize math performance. *Science, 321*, 494–495.

Ioannidis, J. P. (2005). Why most published research findings are false. *PLoS Medicine, 2*, e124.

———. (2008). Why most discovered true associations are inflated. *Epidemiology, 19*, 640–648.

Kerr, N. L. (1998). HARKing: Hypothesizing after the results are known. *Personality and Social Psychology Review, 2*, 196–217.

Kirk, A. P., Jain, S., Pocock, S., Thomas, H. C., & Sherlock, S. (1980). Late results of the Royal Free Hospital prospective controlled trial of prednisolone therapy in hepatitis B surface antigen negative chronic active hepatitis. *Gut, 21*, 78–83.

Kline, R. B. (2004). *Beyond significance testing: Reforming data analysis methods in behavioral research*. New York: American Psychological Association.

Kriegeskorte, N., Simmons, W. K., Bellgowan, P. S. F., & Baker, C. I. (2009). Circular analysis in systems neuroscience: The dangers of double dipping. *Nature Neuroscience, 12*, 535–540.

Lamb, E. L. (2012, July 17). 5 sigma—What's that? *Scientific American*. Retrieved November 16, 2012, from http://blogs.scientificamerican.com/observations/2012/07/17/five-sigmawhats-that/

Larsson, S. C., & Wolk, A. (2006). Meat consumption and risk of colorectal cancer: A meta-analysis of prospective studies. *International Journal of Cancer, 119*, 2657–2664.

Lee, K. L., McNeer, J. F., Starmer, C. F., Harris, P. J., & Rosati, R. A. (1980). Clinical judgment and statistics. Lessons from a simulated randomized trial in coronary artery disease. *Circulation, 61*, 508–515.

Lenth, R. V. (2001). Some practical guidelines for effective sample size determination. *The American Statistician, 55*, 187–193.

Lewin, T. (2008, July 25). Math scores show no gap for girls, study finds. *The New York Times*. Retrieved July 26, 2008, from http://www.nytimes.com/2008/07/25/education/25math.html

Limpert, E., Stahel, W. A., & Abbt, M. (2001). Log-normal distributions across the sciences: Keys and clues. *Biosciences, 51*, 341–352.

Mackowiak, P. A., Wasserman, S. S., & Levine, M. M. (1992). A critical appraisal of 98.6 degrees F, the upper limit of the normal body temperature, and other legacies of Carl Reinhold August Wunderlich. *Journal of the American Medical Association, 268*, 1578–1580.

Macrae, D., Grieve, R., Allen, E., Sadique, Z., Morris, K., Pappachan, J., Parslow, et al. (2014). A randomized trial of hyperglycemic control in pediatric intensive care. *New England Journal of Medicine, 370*, 107–118.

Masicampo, E. J., & Lalande, D. R. (2012). A peculiar prevalence of p values just below .05. *Quarterly Journal of Experimental Psychology, 65*, 2271–2279. doi:10.1080/17470218.2012.711335

Messerili, F. H. (2012). Chocolate consumption, cognitive function, and Nobel laureates. *New England Journal of Medicine, 367*, 1562–1564.

Metcalfe, C. (2011). Holding onto power: Why confidence intervals are not (usually) the best basis for sample size calculations. *Trials, 12*, A101.

Motulsky, H. J. (2014). *Intuitive biostatistics: A nonmathematical introduction to statistical thinking*, 3rd ed. New York: Oxford University Press.

Motulsky, H. J., O'Connor, D. T., & Insel, P. A. (1983). Platelet alpha 2-adrenergic receptors in treated and untreated essential hypertension. *Clinical Science, 64*, 265–272.

Munger, K. L, Levin, L. I., Massa, J., Horst, R., Orban, T., & Ascherio, A. (2013). Preclinical serum 25-hydroxyvitamin D levels and risk of Type 1 diabetes in a cohort of US military personnel. *American Journal of Epidemiology, 177*, 441–419. doi:10.1093/aje/kws243

Nieuwenhuis, S., Forstmann, B. U., & Wagenmakers, E.-J. (2011). Erroneous analyses of interactions in neuroscience: A problem of significance. *Nature Neuroscience, 14*, 1105–1107.

Oremus, W. (2013, June 12). The wedding industry's pricey little secret. *Slate*. Retrieved June 13, 2013, from http://slate.me/1a4KLuH

Parker, R. A., & Berman, N. G. (2003). Sample size: More than calculations. *The American Statistician, 57*, 166–170.

Prinz, F. F., Schlange, T. T., & Asadullah, K. K. (2011). Believe it or not: How much can we rely on published data on potential drug targets? *Nature Reviews Drug Discovery, 10*, 712.

Pukelsheim, F. (1990). Robustness of statistical gossip and the Antarctic ozone hole. *The IMS Bulletin, 19,* 540–545.

Roberts, S. (2004). Self-experimentation as a source of new ideas: Ten examples about sleep, mood, health, and weight. *Behavioral and Brain Sciences, 27,* 227–262; discussion 262–287.

Rothman, K. J. (1990). No adjustments are needed for multiple comparisons. *Epidemiology, 1,* 43–46.

Russo, J. E., & Schoemaker, P. J. H. (1989). *Decision traps. The ten barriers to brilliant decision-asking and how to overcome them.* New York: Simon & Schuster. ISBN: 0671726099.

Samaniego, F. J. A. (2008). Conversation with Myles Hollander. *Statistical Science, 23,* 420–438.

Schoemaker, A. L. (1996). What's normal?—Temperature, gender, and heart rate. *Journal of Statistics Education, 4,* 2. Retrieved May 5, 2007, from http://www.amstat.org/publications/jse/v4n2/datasets.shoemaker.html

Sellke, T., Bayarri, M. J. & Berger, J. O. (2001). Calibration of ρ values for testing precise null hypotheses. *The American Statistician, 55,* 62–71.

Simmons, J. P., Nelson, L. D., & Simonsohn, U. (2011). False-positive psychology: Undisclosed flexibility in data collection and analysis allows presenting anything as significant. *Psychological Science, 22,* 1359–1366.

Simon, S. (2005, May 16). Stats: Standard deviation versus standard error. Retrieved December 20, 2012, from http://www.pmean.com/05/StandardError.html

Spector, R., & Vesell, E. S. (2006). The heart of drug discovery and development: Rational target selection. *Pharmacology, 77,* 85–92.

Squire, P. (1988). Why the 1936 *Literary Digest* poll failed. *Public Opinion Quarterly, 52,* 125–133.

Statwing. (2012, December 20). The ecological fallacy. Retrieved February 8, 2013, from http://blog.statwing.com/the-ecological-fallacy/

Svensson, S., Menkes, D. B., & Lexchin, J. (2013). Surrogate outcomes in clinical trials: a cautionary tale. *JAMA Internal Medicine, 173,* 611–612.

UCSF Legacy Tobacco Documents Library. (2008). Weak associations. Retrieved January 10, 2013, from http://legacy.library.ucsf.edu/tid/rmc80b00/pdf

US Census Bureau. (2011). Median household income by state: 1984 to 2011. Retrieved January 30, 2012, from http://www.census.gov/hhes/www/income/data/historical/household/2011/H08_2011.xls

Vedula, S. S., Bero, L., Scherer, R. W., & Dickersin, K. (2009). Outcome reporting in industry-sponsored trials of gabapentin for off-label use. *New England Journal of Medicine, 361,* 1963–1971.

Velleman, P. F., & Wilkinson, L. (1993). Nominal, ordinal, interval, and ratio typologies are misleading. *The American Statistician, 47,* 65–72.

Vera-Badillo, F. E., Shapiro, R., Ocana, A., Amir, E., & Tannock, I. F. (2013). Bias in reporting of end points of efficacy and toxicity in randomized, clinical trials for women with breast cancer. *Annals of Oncology, 24,* 1238–1244. doi:10.1093/annonc/mds636

Vickers, A. J. (2006, July 26). Shoot first and ask questions later: How to approach statistics like a real clinician. *Medscape Business of Medicine, 7,* 2. Retrieved June 19, 2009, from http://www.medscape.com/viewarticle/540898

———. (2010). *What is a P-value anyway?* Boston: Addison-Wesley. ISBN: 0-321-62930-2.

von Hippel, P.T. (2005). Mean, median, and skew: Correcting a textbook rule. *Journal of Statistics Education, 13*(2).

Wasserman, L. (2012, August 16). P values gone wild and multiscale madness. Retrieved December 10, 2012, from http://normaldeviate.wordpress.com/2012/08/16/p-values-gone-wild-and-multiscale-madness/

Ziliak, S., & McCloskey, D. N. (2008). *The cult of statistical significance: How the standard error costs us jobs, justice, and lives.* Ann Arbor: University of Michigan Press. ISBN: 0472050079.

INDEX

Printed in the USA/Agawam, MA
July 19, 2023

813123.036